Clean Technology for the Manufacture of Speciality Chemicals

Clean Technology for the Manufacture of Speciality Chemicals

Edited by

W. Hoyle
Consultant

M. Lancaster
University of York, UK

RS•C

ROYAL SOCIETY OF CHEMISTRY

The proceedings of the symposium Clean Technology for the Manufacture of Speciality Chemicals held on 26 and 27 September 2000 at New Century House, Manchester, UK

Special Publication No. 260

ISBN 0-85404-885-5

A catalogue record for this book is available from the British Library

Published by The Royal Society of Chemistry,
Thomas Graham House, Science Park, Milton Road,
Cambridge CB4 0WF, UK

Registered Charity No. 207890

For further information see our web site at www.rsc.org

Printed and bound in Great Britain by Bookcraft (Bath) Ltd.

Preface

It is interesting to look at the major sources of waste from within the chemical industry. Bulk chemical production is normally thought of as a 'dirty' industry and historically this has been the case. As the industry became more competitive, more economic processes needed to be developed and it became unacceptable (and non-viable) to produce the same amount of waste as 30 years ago. Today it is the speciality and pharmaceutical industries which produce most waste (at least on a relative basis). Although the situation is slowly changing, it is still the case, in these high added value industries, that the cost of manufacture is low compared to the selling price of the product. It is these industry sectors where Clean Technology and Green Chemistry can produce the most beneficial effects.

Manufacturing of speciality chemicals is becoming more clean and sustainable due to a combination of improved technology, increased environmental awareness and the realisation that waste minimisation at source has significant bottom line advantages. This is the focus of the first paper, which highlights the cost of waste to the speciality chemicals sector. Industry is increasingly considering the triple bottom line and, as financial accounting is becoming more focussed at the individual process level, it is becoming easier to identify and justify process development work on the ground that clean production does pay.

A vision of the future, in which industry only makes the specific molecules wanted by customers in small, intensified processes, is presented in the second paper. How we get there is likely to be a combination of incremental improvements plus the occasional radical leap forward through development of new technology. Both of these approaches will be enhanced by the sharing of best practice within the sector; specific examples and the generalities of this approach are presented in the third paper. In many instances it is obvious that a piece of research will lead to a more eco-efficient process, but this is not always the case. Currently there is no agreed measure of how 'green' a process is; this aspect is covered in the next paper which gives a unique insight into how a leading pharmaceutical company has developed green metrics.

Chemical engineering has a large role to play in greening process the next two papers cover the subject of process intensification in some detail, presenting industry case studies of what can be achieved. The next two papers present case studies of how the generic oxidation and nitration reactions can be significantly greened. These papers are followed by a discussion of the role of catalysis in the speciality chemicals industry. Several real examples are presented which show the scope for transforming the industry to provide similar benefits to those obtained in the bulk chemical industry. Whilst the focus of the book has been on the front end of the process, inevitably end of pipe treatment is necessary in some cases. The final paper discusses the common problem of solvent recycling with a case study of successful recycling of complex solvent mixtures.

We are very grateful to all the contributors for giving so generously of their valuable time and making it possible to produce this book.

Mike Lancaster
Bill Hoyle

Contents

CLEAN TECHNOLOGY FOR SPECIALITY CHEMICALS

Mike Lancaster

Manager Green Chemistry Network
Department of Chemistry
University of York
York YO10 5DD

1 INTRODUCTION

The concepts of clean technology and green chemistry have been around for 10 years or so and slowly but surely these ideas are gaining acceptance by a wider cross section of industry. One of the main obstacles to progress is the mistaken belief that clean technology, which is designed to limit pollution and environmental damage, equates to a reduction in bottom line profitability. Can a modern speciality chemical business can be successfully run on the principles of the Triple Bottom Line and are Clean Technology and increased profitability happy bedfellows? This is a question many people are starting to ask, and as the papers in this book will demonstrate the answer is a resounding yes.

So what is Clean Technology or Green Chemistry all about? First and foremost it is about reducing waste. Waste is increasingly expensive to dispose of and the major source of pollution coming from the chemical industry. Maximising atom efficiency is linked to waste reduction. The concept of designing chemical reactions such that as many atoms of starting material as possible end up in useful product may seem common sense but sadly is not often followed. Secondly it is about reducing materials, this includes raw materials, and materials of construction; this area brings in the concepts of process intensification which will be fully discussed in later chapters.

Clean technology is also about reducing the hazard and the risk both to people and the environment; this brings in concepts such as inherently safe design of reactions and substitution of hazardous chemicals or those that pose a high risk. In short green chemistry is about reducing the environmental impact of both processes and products.

One important aspect of green chemistry, which is often forgotten, is cost. If industry is to take up these wonderful ideals of green chemistry it will only do so if it makes economic sense. However if we have reduced waste, materials, energy etc. then it is also very likely that the cost has also been reduced (Figure 1).

From the above description it should be evident that Clean technology involves both chemistry and chemical engineering. If industry is to develop the cleaner processes now being demanded we need to establish multidisciplinary teams at the conceptual stage.

Industry need not look on environmental protection as an additional burden imposed by Governments and society it should look at it as an additional opportunity to develop more cost effective processes and products. With a little thought, and perhaps a culture change, the chemical industry can be competitive and environmentally benign.

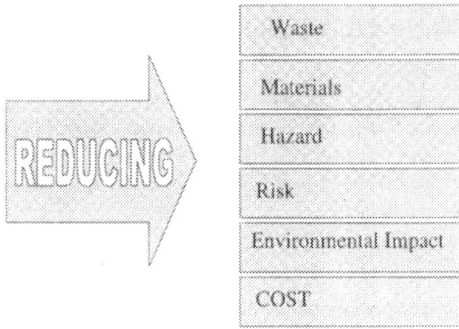

Figure 1 *What clean technology is about*

UK chemical industry expenditure on environmental technology during 1997 was almost a quarter of that spent by UK industry as a whole. Expenditure is divided into 3 - by far the largest is operating expenditure some 65% of the total of over £1,000 million. The rest is capex: some 26% was spent on end of pipe technology with only 9% being spent on what might be considered integrated process clean technology.

Spending on environmental technology almost doubled between 1994 and 1997. It is little wonder that looking at these figures that industry believes that protecting the environment is adding to its un-competitiveness, since the figure spent is over 5% of revenue generated by sales. However I would like to present these figures as an opportunity rather than a threat to the industry. Only £100 million pounds is currently spent on what I would consider the most valuable area of clean technology. The real opportunity lies in increasing this expenditure by a factor of 2 or 3 and eliminating a significant part of this end of pipe and operating costs (Figure 2).

Chemicals Sector Expenditure on Environmental Control (1997)

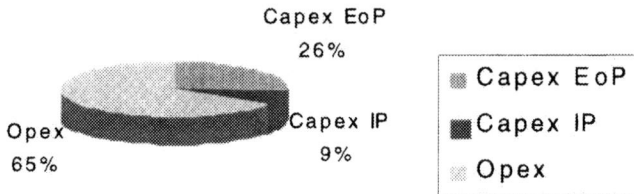

1997 Total £1042 m

1994 Total £547 m

Figure 2 *Expenditure on environmental control*

The root cause of this vast sum spent on environmental technology is waste, and as Sheldon[1] has pointed out it is the fine, speciality and pharmaceutical sectors which have more of a problem with waste (on a Kg per Kg product basis) than the bulk chemicals industry. Looking closely at the cost of waste within the speciality chemicals sector is a little more difficult. For many smaller companies working with multipurpose plants the true breakdown of manufacturing costs is still often unknown - overheads can be used to hide a multitude of sins! With the rapid development in sophisticated process monitoring equipment and state of the art control systems the true production cost is slowly becoming evident. The most important information coming out of this analysis is that the cost of waste, (including effluent treatment, waste disposal, loss of raw materials etc.) often amounts to some 40% of the overall production costs (Figure 3).

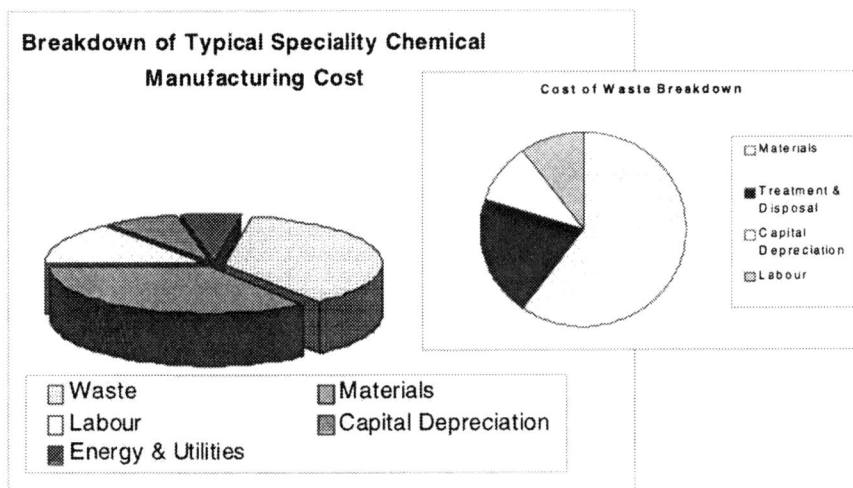

Breakdown of Typical Speciality Chemical Manufacturing Cost

Cost of Waste Breakdown

- Materials
- Treatment & Disposal
- Capital Depreciation
- Labour

- ☐ Waste
- ☐ Labour
- ■ Energy & Utilities
- ■ Materials
- ▨ Capital Depreciation

Figure 3 *Breakdown of the cost of waste*

So what are the technologies that are being used or are likely to be used in the near future to both drive down costs and create more environmentally friendly processes? The bulk chemicals industry has been transformed by the increased use of catalysis

Today there are around 130 chemical manufacturing processes such as alkylation, isomerisation, amination and etherification using catalysts such as zeolites, ion exchange resins, clays and complex oxides. This technology has largely been developed to maintain competitiveness, improve product quality, and improve process efficiency. For example the waste produced from the very large scale use of aluminium chloride in alkylation reactions started to make the process uneconomic. Introduction of zeolites and other heterogeneous catalysts has had both significant cost and environmental benefits. It is untrue to say that catalysts are not used in the speciality chemicals industry but their use is not as widespread as it could be. Potential cost benefits result from faster processes, higher selectivities and lower energy use (Figure 4).

Supercritical fluid technology has been around for many years but, largely due to the relatively high pressures involved, remained a laboratory curiosity. Supercritical carbon dioxide is commercially used for extraction processes and the technology is now being

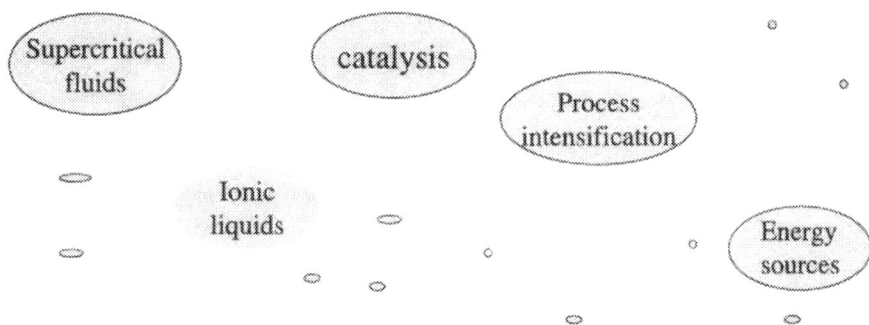

Figure 4 *Some newer clean technologies*

looked at seriously for chemicals manufacture. The reasons for this are that improvements in engineering mean that the equipment is more affordable and when combined with modern catalyst technology the rates of some reactions are very fast. Hence for true low volume specialities there is the potential for producing requirements from a small relatively low cost plant. Companies such as Thomas Swan and Hoffman La Roche are now looking to commercialise the technology for reactions like continuous hydrogenation and alkylation. It is not a universal answer but is likely to provide true competitive advantage in a niche market.

Ionic liquids are another emerging technology which may avoid the use of organic solvents. Although these materials are strong acids and may not be particularly environmentally friendly in their own right they can be readily recycled due to their immiscibility with most organics. They are being studied particularly for reactions which require highly acidic conditions such as alkylation where they act as both a solvent and a catalyst.

Process intensification is one of the chemical engineering solutions to clean technology. The main benefits are often lower energy uses, lower capital costs and increased throughput from a smaller plant derived from different engineering conceptual designs. The next generation of plants will be smaller, cheaper, and more environmentally friendly to run.

As the energy intensity of a process becomes more of an issue on cost and environmental grounds we shall start to see processes being developed using microwave, ultrasonic, electrochemical and photochemical reactors. For some reactions these techniques will not only deliver energy efficiency but may also lead to higher selectivities and atom efficiencies.

Environmental concerns are here to stay. They can be viewed as a threat requiring ever increasing expenditure on end of pipe technologies to meet ever-increasing legislation or they can be viewed as an opportunity to introduce cleaner processes, which are more efficient and cost effective.

References

1.   R. A. Sheldon, *Chemtech*, 1994, March, 38.

CLEAN TECHNOLOGIES TO MEET ECONOMIC, ENVIRONMENTAL AND SAFETY NEEDS

D. S. Ridley

Environment Agency
Coverdale House, Aviator Court
Amy Johnson House, Amy Johnson Way
Clifton Moor
York YO30 4GZ

1 INTRODUCTION

The objective of this paper is to consider the pressures on today's industry, to identify the commonality between these pressures and to identify the solutions for the future. Examples will be used to illustrate that to succeed in the future our thinking and approach towards science and technology will have to be radically different to that of today.

The first question to be asked before considering the role of the chemical industry in the future is to question its need. The world's population continues to increase from today's all time maximum. The life expectancy of both the fit and the ill in the developed world continues to increase. Our response to natural disasters or problems threatening life anywhere around the world invariably involves the deployment of the products of the worlds chemical industry. The developing world progresses into the developed world by acquiring or producing amongst other products the products of the chemical industry.

Quality of life is a complex and controversial parameter to define. Suffice to say that the majority of definitions of improvement in a quality of life directly or indirectly refer to the availability or use of the products of the Chemical Industry. It is inconceivable that our current quality of life let alone that of the future could be sustained without a viable chemical industry. The above analysis should not be interpreted as inferring that the industry has some form of divine right to its existence, in order to be viable the chemical industry needs to constantly adapt and change to meet the needs of society.

Looking ahead to what will be important for the industry in the future, emphasis is likely to be placed on the following aspects:-

- Product quality and efficacy
- Competitive price
- Return on investment
- Exciting place to work
- Being heard but not seen

What is proposed as important in the future is no different to what will be construed by many as being important today. What may surprise is the likely extent and potential pace of the necessary changes. The pace of change is likely to be set by the rate of development of the already developing eastern economies. In the main these economies are well endowed with creative technical expertise and as they expand they are installing technology at least the match of our own.

2 PRODUCT QUALITY AND EFFICACY

Product quality and efficacy improvements are likely to be driven more from the litigation based developed nations, than by eastern competition, at least in the short-term. In the fine chemical industry the need to produce and isolate only the precise isomer of the active molecule is likely to increase i.e. we need to make the exact molecule we want. Whilst new chemistry, new catalysts and new extractive routes will all play their part so also will the need to precisely define and control the reactions and extractions at a level not currently available and in many instances not yet dreamed of. In the polymer industries, in common with other performance or property products, the ability to produce the specific product for the specific purpose will provide the more lucrative markets.

In short meeting the individual customers individual needs, whilst possibly assisting the customer to identify and specify their specific need is likely to provide the more lucrative and sustainable markets. Not only does this raise the spectre of new or much refined existing technologies it also questions the economic scale of a future plant. Does the engineering need to be developed to ensure that a plant sized to provide a customer's precise needs is economic rather than having to scale a plant to serve the entire markets need before it can be judged to be economic.

3 COMPETITIVE PRICE

The premium payment for supplying a customised or specific product to a customer to allow them to optimise or customise their product should continue. Additionally the concept of sustainability is beginning to take root. Questions at annual general meetings are beginning to include pertinent questions relating to the environmental footprint and the social conscience of the organisation its suppliers and contractors. The Chemical industry is possibly disproportionately blamed for its perceived environmental impact. All companies express their economic performance as a series of factors all expressed in or easily convertible into US Dollars. Surely we need to develop and agree, with our stakeholders, the methods and means by which we can calculate an organisations environmental and social balance sheet in order to confirm and reply to or alternatively deny the charges levelled at almost all organisations within or related to the chemical industry. An example of such an environmental tool is given later.

4 RETURN ON INVESTMENT

In the recent past the capital value of the physical plant normally dominated the balance sheet. Today a number of companies have to include a greater provision to cover the potential costs of remedying problems or liabilities, associated with the plant such as land contamination or problems associated with either the use or abuse of the product or by-products. Whilst in certain well-publicised cases these costs are indeed very substantial, many other companies are blighted by the possibility of the costs. Much modern regulation, such as Control of Major Accident Hazards (COMAH), Integrated Pollution

Prevention and Control (IPPC), Integrated Pollution Control (IPC) and much of the waste legislation are rightly more about liability management and minimisation than about direct impact. Whilst I acknowledge that a banker or insurer's greatest hate is an unquantifiable risk and that the greatest joy of a prosecuting lawyer is to use a risk assessment as proof of the company's willingness to take risks. I believe that the majority, including a presiding judge, realise that progress is only made through the management of risk which has at its core 'do the benefits outweigh the dangers'. In view of this the 'applications for permitting' under the above legislation should not only assess and quantify the risks of the process or product but should also quantify the benefits. To do this in an uncontentious manner will require the development of a new system of benefit assessment, yet another new tool for the future.

The minimisation of capital has exercised the minds of every generation of engineers, so what is new today. In spite of several years of low and relatively predictable inflation and interest rates the financial markets expectation is often for double figure growth at minimum capital investment. The hapless Chemical Company finance director seeking several hundreds of millions of pounds over 20 years is unlikely to excite the market.

The availability of grants has influenced the spectrum of investment decisions over recent decades. It is important to recognise that the European support systems are likely to focus more on rebuilding the industrial infrastructure of the new eastern member or applicant states than on that of the more established economies.

5 EXCITING PLACE TO WORK

The success of the industry is not only dependent upon the supply of sufficient trained staff to design, build, operate and maintain the process, but upon an adequate supply of these trained individuals into all the associated businesses needed to support the primary business. These businesses include banking, insurance, regulation and the media. Without individuals in these associated businesses with a comprehension of the technologies, the risks and benefits of the processes a balanced representation of the viability, risks and benefits offered by a specific proposal or plant will be so much harder to gain.

As a result of the above it is essential to retain a greater number of students beyond GCSE than is currently being achieved. To do this some curricula for the study of sciences, engineering and chemistry in particular may need rejuvenating to make them more dynamic and pertinent to modern life.

6 BE HEARD BUT NOT SEEN

The chemical industry is frequently portrayed in a rather negative manner. Many individuals can not relate the medicine they are taking, the fabric they are wearing, the abundant clean food they are consuming or the cleaner fuel they are using to the chemical industry. In the more extreme situations the plant passed on the way to work is a place of mystery and suspicion. With few cars in the car park and protected by an aggressive fence all that is produced is the occasional bad smells or noise. No useful product emerges. The positive initiatives taken by the industry or the leading edge techniques used to assess issues such as risk are not sufficiently publicised or pushed. These associations are necessary if we are to begin to change the extreme but not unrealistically negative image of the industry outlined above. Increasingly industry is responding to societies demand for zero emissions, no hazards and no accidents through initiatives such as responsible care, cradle to grave responsibility and risk and liability management.

7 THE CHEMISTRY OF THE FUTURE

The above is something of a wish list for a competitive, hazard and pollution free chemical industry but the question is can this be achieved with current technology in a standard 500 gallon glass lined stirred reactor? The answer is surely no, chemistry of the future will require:

- Intense controlled reaction conditions
- Intense controlled energy transfer
- Precise and rapid control of all relevant reaction parameters

By having high intensity reactors we will have lower inventories, reduced hazards and lower capital costs. Hence the reactor of the future is likely to be the corollary of today's design. It is this challenge which will provide the challenge and excitement needed to keep the industry vibrant. The requirements for a competitive, hazard free eco-friendly industry are not contradictory. What is essential is to think broad enough to see the solution. Later in the book some excellent work is described showing that a nitration reaction could be made without using sulphuric acid. Extrapolating, the intention of this project asks if the nitration could be done without nitric acid. The acids need to be removed as it is the accompanying water that initially forms the aqueous discharge from the process which costs so much to clean before its return to nature. To achieve this would appear theoretically possible. To do it would require catalysts; ion concentration control or electrode potential control hitherto regarded as impractical. But if you start by thinking that the reactor will resemble a fuel cell rather than a 500-gallon glass vessel, then the technology may become practical. At the same time the inventory, scale problems and associated costs are revolutionised. None of the above is known fact. It is this radical re-engineering of our processes which may prove essential in building a sustainable UK Industry.

8 THE PLANT OF THE FUTURE

The plant of the future may well look considerably different to that of today. Whilst not wishing to appear dismissive of either the quantity or quality of information many organisations or individuals hold about the processes they either design or operate, there are some where we clearly do not know or understand the mechanism and mechanics of the process to the depth or refinement implied above. It is by understanding the processes at the micro level that we are best placed to identify the higher intensity route to producing the sole desired product. It is this higher intensity precisely controlled process which offers the prospect of significantly reducing the scale of a plant and hence the associated costs, inventory and liabilities. We need to know:

- Precisely how it works
- The risks & how to manage them
- The impacts and how to manage them

By way of example to bring meaning to some of the concepts and opinions expressed above, a development study using a technique currently being developed by the author is described for a modern municipal waste incinerator.[†] To operate not only must waste be burnt, but waste must be transported to the incinerator and ash transported away. The incinerator also generates some electricity from the waste heat; this is supplied into the grid.

Figure 1, shows an edited version of the pollution inventory for the process. This data is accessible on the net. The data represents the number of Kg of each pollutant released over the course of the specific year for which the data was reported. The conclusion drawn is that this raw data conveys little about the environmental significance of the process.

NOx = 197142 kg
PM10 = 11624 kg
HCl = 44859 kg
Cd = 15.66 kg
TEQ= 0.17 grams

Figure 1 *Pollution inventory for waste incinerator*[1]

Figure 2 shows the output following processing the raw data from Figure 1 through the model, which calculates the process impact as a single factor and then calculates what percentage each pollutant contributes to the total. This infers that HCl; NOx and cadmium should be the order of environmental concern.

NOx = 24.4%
PM10 = 1.5%
HCl = 33.5 %
Cd = 16.4 %
TEQ = 8.9 %

Figure 2 *Process outputs for waste incinerator*[1]

[†] This does not represent or infer Environment Agency Policy.
1. PM10 is the particulate matter less than 10 microns in diameter, and TEQ is the Toxic Equivalent Quotient.

Figure 3 shows the total impact of the incineration process compared to that of the incurred transport. This could be used to determine the environmentally viable or optimal size of the process.

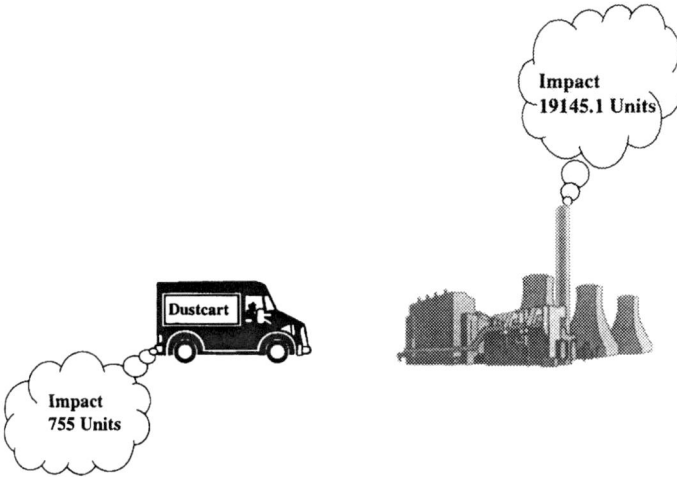

Figure 3 *Impact of incineration for incinerator plus transport*

Figure 4 continues the theme and includes the environmental credit for the electricity generated. It is the analysis of this Figure that gives the real process insight. If the desire were to cut the impact of the process, then beyond the issues of Figure 2 the most fruitful route would be to increase the energy efficiency. This is constrained by the low steam temperature and a use for around 80 MW of warm water.

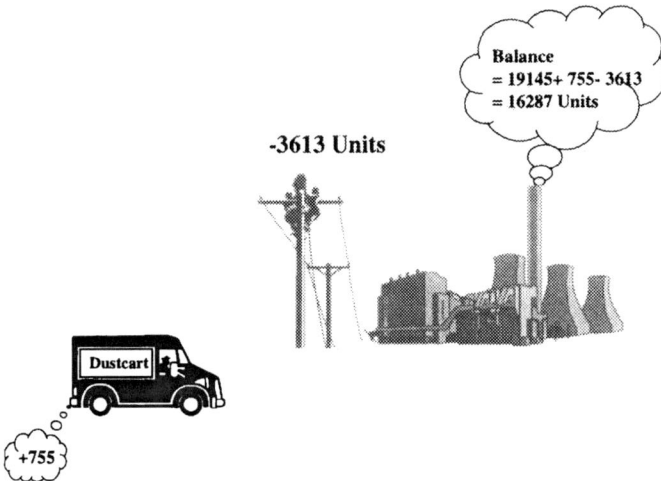

Figure 4 *The real environmental cost for incinerator, transport and power credit*

A solution may be to build a unit 1/1000 the size of this. Clearly the original combustion process would not be economically viable. Using alternative thermal processes a greater percentage of electricity could be generated with the waste heat used even to heat a building. In short a dramatic environmental win. Socially the picture is very different to the norm. In that in place of asking a small community to be host to a large plant serving a very large hinterland. We are now asking a small community to be host to a small plant serving them and whose by-product heat and electricity could possibly be used for their benefit or gain. The technical approach and challenge is no less radical. For every one current incinerator we would need 1000 of these, they must be as clean as the original, they must be simple to operate, they must be reliable, they must monitor and manage themselves and finally they must be price competitive.

By using this type of model, identifying opportunities, and then thinking radically about how the challenges can be met and satisfied economically, socially and technically, have we identified the viable sustainable future.

THE CLEAN TECHNOLOGY ROUTE TO WASTE MINIMISATION

Adrian J. Cole

Environmental Technology Best Practice Programme
AEA Technology Environment
Harwell
Didcot
Oxon OX11 0RA

1 INTRODUCTION

Companies are increasingly recognising the need for environmental improvement – and also seeing the financial benefits that can be achieved. The stimulus to embark on an improvement programme can come from many sources – deciding to implement an Environmental Management System, a need to meet customer specifications and changes in the regulatory requirements are all common drivers. The chemical industry has often been at the leading edge in seeking ways to meet the challenges of environmental improvement. This is perhaps not surprising given the environmental risks that are associated with many of the industry's raw materials, processes and products and the consequent level of regulation that the industry is subject to.

The Environmental Technology Best Practice Programme (ETBPP) is a Government programme that aims to help UK industry and commerce to make effective environmental improvements, whilst also saving money. The programme has been working with the speciality chemicals sector for over three years and this paper uses some of the case studies that have been published to illustrate the varying routes that companies have adopted to realise cost-effective environmental improvement. Finally, the wide range of information and guidance that is available to companies from the Programme is outlined.

2 WASTE MINIMISATION OR CLEAN TECHNOLOGY?

So how should a company go about improving its environmental performance? Depending on the nature of the proposed actions, the routes companies can take are often labelled as Waste Minimisation approaches or Clean Technology approaches. Commonly used descriptions of these are:

Waste Minimisation: management procedures to prevent, reduce or re-use waste material streams - *does not usually involve investing in major plant*
Cleaner Technology: 'step' changes to prevent, reduce or re-use waste material streams - *often involves investing in major new plant.*

This duality is reflected in the ETBPP, which has two permanent themes of waste minimisation and clean technology as core aspect of its activities. However, giving

Table 1 *Perceptions of Waste Minimisation and Clean Technology Approaches*

Waste Minimisation	*Clean Technology*
small benefits?	large benefits?
effort intensive?	capital intensive?
limited scope?	high profile?
boring?	exciting?

techniques labels in this way can imply that the approaches are mutually exclusive to each other and lead to other perceptions – see Table 1 - of their relative value.

Technology based companies, such as those in the chemicals sector, may have a tendency to favour Clean Technology approaches, operating as many do in a culture of new product cycles driving major plant or process modifications. Waste minimisation techniques can be seen as both too difficult because of the emphasis on changes in management and operating practices, and at the same time, too simple to achieve the required benefits! This dichotomy of approach is illustrated in Figure 1 a-f.

The company at the bottom left of the charts has decided that it wants to improve its performance and move up to the top right, over a suitable period of time. Of course it is operating in compliance with the current regulatory requirements – as all good companies do! The vertical axis of the charts has not been defined. It represents some combined algorithm of performance in environmental, health and safety and operational efficiency and has been labelled "quality" for convenience. Current fashion might suggest it should be labelled as a "sustainability index". The company could implement a vigorous waste minimisation campaign to achieve its goal, as shown in Figure 1a. This could involve, for example

- Creating improvement teams across the site.
- Assessing each site activity to identify wastes and ways of eliminating, reducing or resusing the wastes.
- Training operating staff in new procedures.
- Setting up a monitoring programme to show the effects of the programme.
- Establishing continual improvement targets for all aspects of the companies operations; and reporting progress to the senior management.

It also means keeping all these activities going over an extended period of time whilst also achieving normal business targets for turnover and profit. Clearly there is a risk of the momentum not being maintained and the company not achieving its goal, as shown in Figure 1b.

Worried by this risk of failure the company might decide that a better approach would be to invest in a new production process to achieve the goal in a single step change, as shown in Figure 1c. The new plant could be developed off-line ensuring minimal disruption to the current production and revenue stream, and involve only a small team and outside suppliers. The new range of "green" products will carry a market premium to justify the significant capital costs. Unfortunately, this approach also has risks as shown in Figure 1d. New technologies can often take longer to perfect and implement than originally expected. They might even fail completely due to some unforeseen problem.

To make matters worse, it is more than likely that a major factor behind the company's plan to improve its performance was the strong possibility of an increase in the regulatory standards applied to the industry – see Figure 1e.

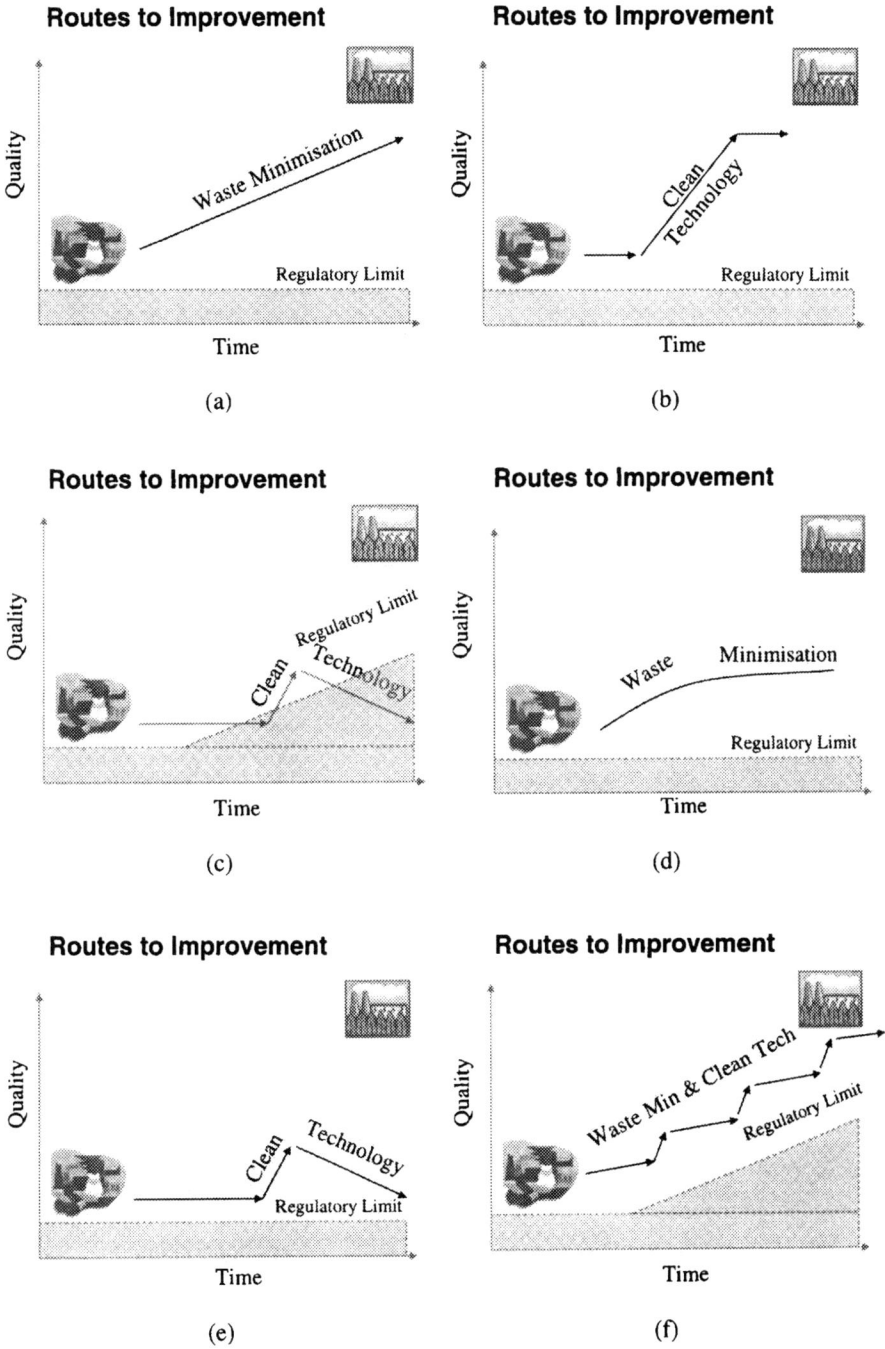

Figure 1 *Routes to environmental improvement*

Clearly an approach more likely to achieve the company's goal is to combine clean technology changes into an underpinning waste minimisation programme, as shown in Figure 1f. Improving the company's performance by a mix of management measures to reduce wastes and limited scale investment in cleaner and more efficient plant can help put the company on a sounder footing to meet its obligations to the regulators, its investors and its staff. It can then plan for the major cleaner technology investment opportunities with less chance of damaging failure.

3 STEPS TO IMPLEMENT CLEAN TECHNOLOGY SUCCESSFULLY

In some circumstances a waste minimisation approach cannot fully meet a company's needs for environmental improvement. An example would be when an unavoidable waste needs to be reduced in order to comply with a significantly reduced release consent. Following a structured step by step method to planning the solution can enhance the success of implementing a clean technology approach, as illustrated in Table 2.

4 COMPANY EXPERIENCES

The ETBBP programme has worked with a wide range of companies to collect and promote examples of good practice in cost effective environmental improvement. In the chemicals sector case study examples illustrate the range of approaches covering the full spectrum of basic waste minimisation through to high investment in clean technology, with many showing the benefits of combining both approaches. A selection of examples from these case studies is outlined briefly below.

Demonstrating that manufacturing companies can benefit from a well planned and implemented waste minimisation programme is the experience of Merck Ltd, one of the programme's first case studies (GC16), at its chemical reagents manufacturing site and offices.

With the aim of reducing the volume of wastes going to landfill by at least 50%, the company initiative minimised the use of office supplies, established re-use of steel drums and recycling of aluminium and stainless steel. Overall a 78% reduction of waste going to landfill was achieved with a net saving of over £23,000/yr. The lesson here is that any company can make valuable improvements irrespective of the nature of their business. Even speciality chemical manufacturers have office and goods handling functions where a similar approach can be applied.

Another early case study is that of Mold Hygiene Chemicals Company Ltd (GC20), a manufacturer of cleansing and industrial hygiene chemical products.

Stimulated by a significant increase in their effluent disposal costs, the company carried out a comprehensive review of its operations. The bleach product filling system was identified as a major source of waste (requiring neutralising of rinse before discharge as trade effluent) as a result of residues left between batches. By optimising the production scheduling of the range of bleach products this waste was minimised. Similarly, acid wastes were minimised by separating residues and reusing in subsequent batches. As a result the company was able to cut its trade effluent charges by over 40%.

Ciba Speciality Chemicals realised that their existing vacuum system was excessively wasteful in terms of lost product, maintenance requirements and disposal of contaminated lubricating oil (GC235).

Table 2 *Steps to Implement Clean Technology*

STEP 1 LAYING THE FOUNDATIONS FOR CLEANER TECHNOLOGY		
A) Raising Awareness	⇒	Communication, Training, Motivation Other Benefits: quality, market share, avoided costs
B) Recognising Opportunities	⇒	Policy Responsibility Legislation EMS
C) Making the Decision	⇒	Environmental Accounting Making the Case

STEP 2 DESIGNING CLEANER PRODUCTS AND PROCESSES		
D) Cleaner Product Design	⇒	Materials Selection Minimising Quantity End Of Life Manufacturing Steps Manufacturing Process Life Cycle Analysis
E) Alternative Processes	⇒	Catalysts Membranes Supercritical Fluids Ultrasonics Electroprocessing Biotechnology
F) Optimising Processes	⇒	Process Intensification Water Pinch Energy Pinch Updated Plant

STEP 3 RUNNING A CLEANER OPERATION		
G) Optimising Control	⇒	Sensors Monitoring Numerical Control
H) Recovering Raw Materials	⇒	Segregation of Streams Separation Technologies
I) Recovering Solvents/Gases	⇒	Separation Technologies Solvents Membranes Oxidation Electrotechnologies Biotechnology

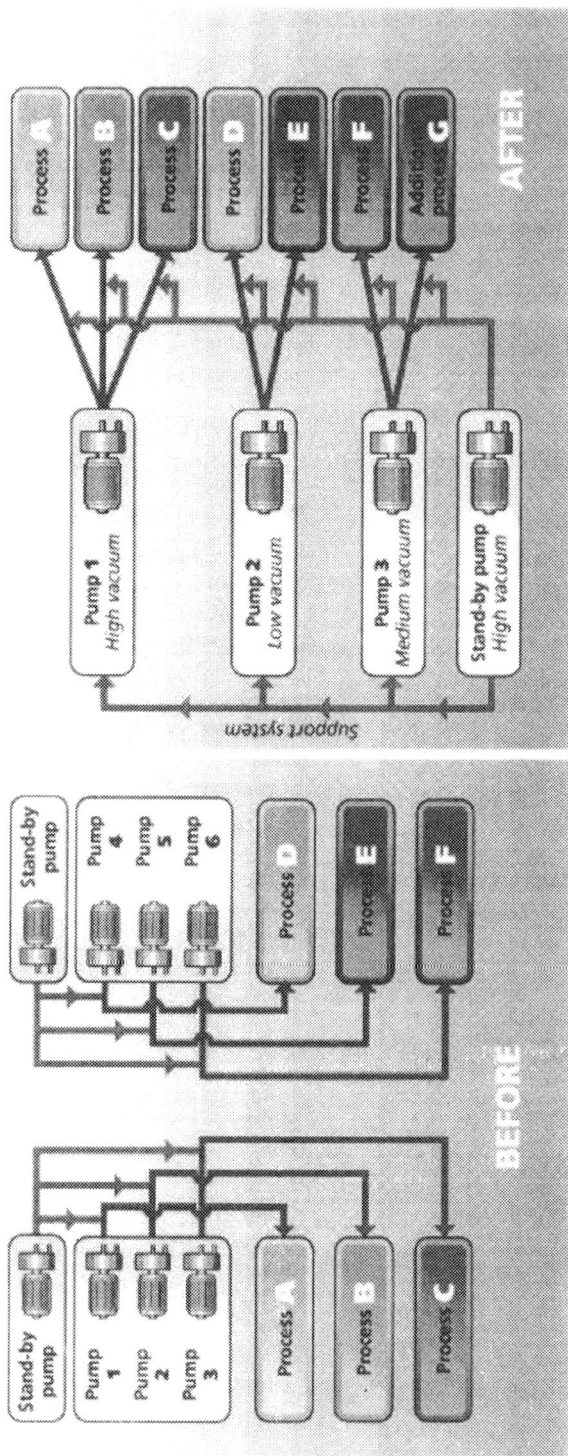

Figure 2 *Rationalisation of vacuum pump arrangement*

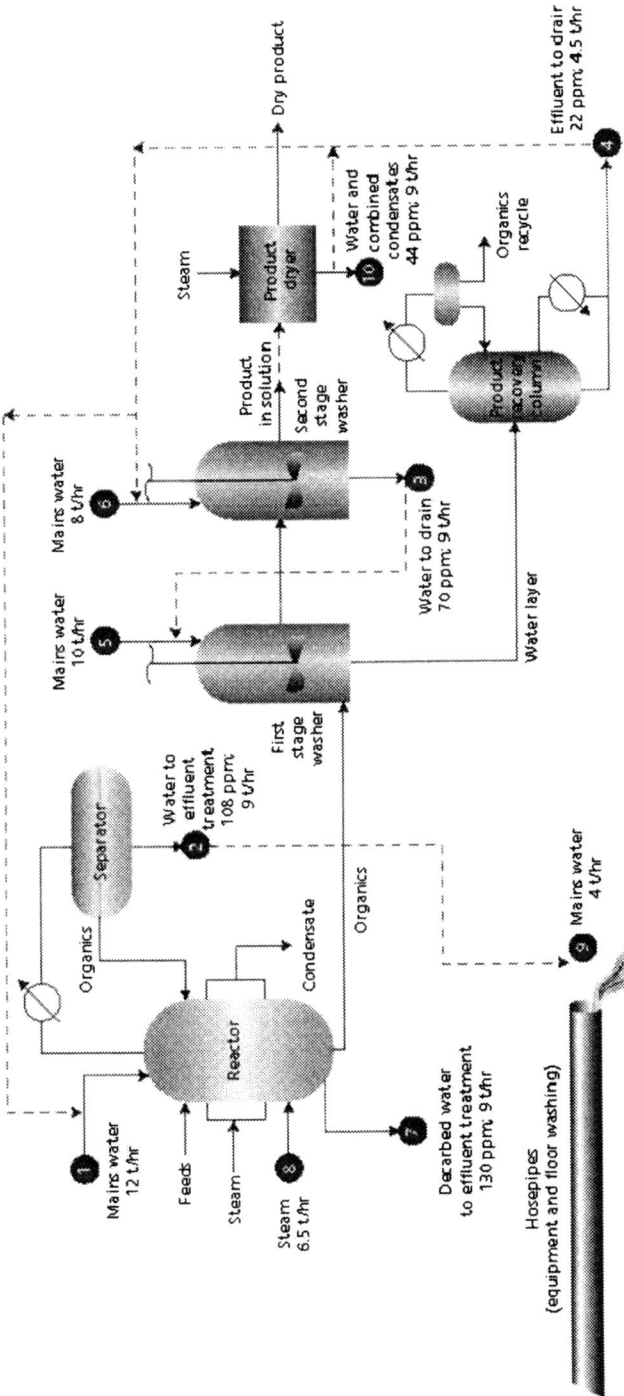

Figure 3 *Illustrative schematic of site water and effluent flows (not actual site)*

Effluent	Flow Rate	Measured Concentration
7	9 t/hr	130 ppm
2	9 t/hr	108 ppm
3	9 t/hr	70 ppm
10	9 t/hr	44 ppm
4	4.5 t/hr	22 ppm

Input	Flow Rate	Max Tolerated Concentration
5	10 t/hr	140 ppm
9	4 t/hr	130 ppm
1	12 t/hr	63 ppm
6	8 t/hr	63 ppm
8	6.5 t/hr	46 ppm

(a)

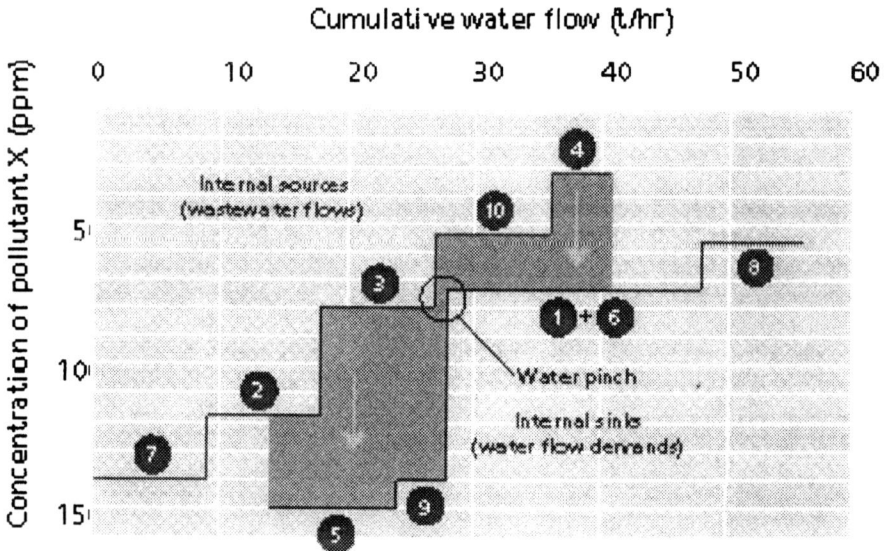

(b)

Figure 4 *Pollutant concentrations and flow rates from Figure 3, listed in descending order of concentration (a), and the resulting graphical representation (b)*

The company was able to make significant improvements by both switching to dry vacuum technology to achieve a 50% reduction in the volume of lubricating oils requiring disposal, and also rationalising the vacuum system to reduce the number of pumps – see Figure 2. The company made savings of over £120,000/yr directly as a result of the waste minimisation actions but also realised production benefits worth over £400,000/yr as result of the reduced maintenance requirements.

Another example of a major chemical manufacturer that has combined waste minimisation and clean technology approaches is Solutia (formerly Monsanto plc, NC55).

With a site water consumption of over 1 million m³/yr, costing over £750,000, and impending stricter consent limits on their discharges, the company was faced with the need to make a significant capital investment in a new effluent treatment plant. Before embarking on this course of action the company carried out a waste minimisation study of their operations which identified opportunities for significant water and raw material savings. This led to the company applying water pinch analysis techniques to its operations to identify even greater savings and negate the need for the new effluent treatment plant, see Figure 3 and Figure 4. This resulted in three reduced scale batch plant modifications being identified for a capital cost of only 10% of that needed for the full effluent treatment plant option. As result water consumption was cut by nearly 50% and annual cost savings of over £800,000 achieved. This illustrates very effectively the benefits of a step by step approach combining waste minimisation and clean technology as outlined earlier.

The final example is of the development and implementation of a fundamental clean technology solution to recover valuable products and solve a waste disposal problem at Elf Atochem UK Ltd (FR200). Using Extractive Membrane Bioreactor technology, see Figure 5, and working with their technology supplier (Membrane Technology Extraction Ltd), the company has developed an easy to use technology for cleaning aluminium trichloride solution contaminated with benzene, toluene and other organic compounds.

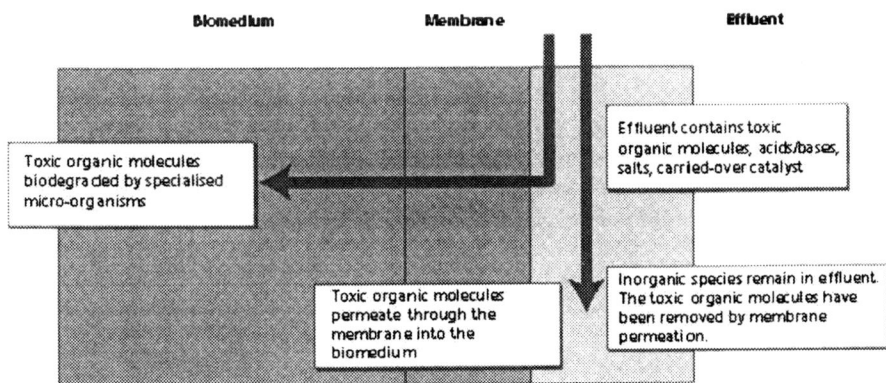

Figure 5 *Operating principle of extractive membrane bioreactor technology*

Benzene removal levels of 99.9%, see Figure 6, have been achieved with an annualised cost of about 10% of conventional off-site disposal. In addition the recovered aluminium trichloride can be sold as product improving the economics even further.

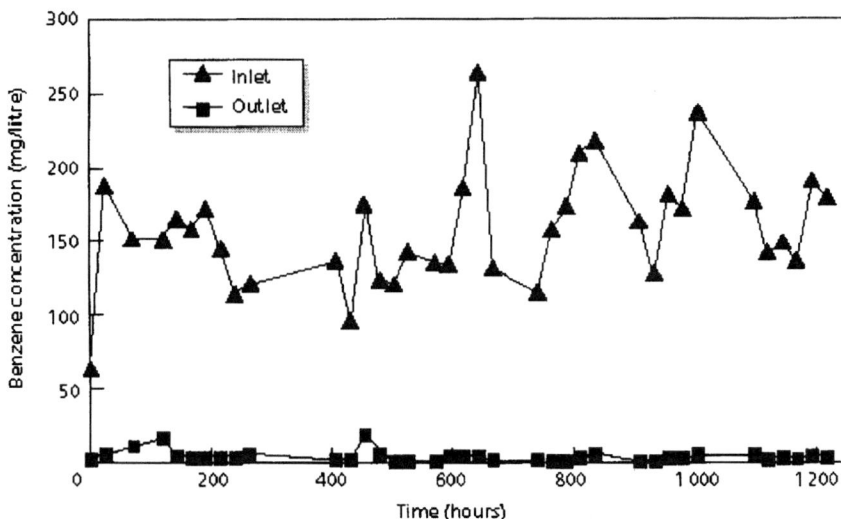

Figure 6 *Measured benzene concentrations*

5 HOW THE ETBPP CAN HELP

In addition to the case studies mentioned above (and several others covering other aspects of environmental improvement such as environmental management systems), the programme has published a range of guides and other tools to help chemical companies. These include:

- Water Use in the Manufacture of Speciality Chemicals (EG105), a benchmarking study to help companies identify water reduction opportunities,
- ERA2000, a software tool, has been developed to help chemical companies benchmark their performance across a range of process and management environmental indicators,
- Reducing Vacuum Costs (GG101), gives guidance on improving the wide range of vacuum systems typically employed in the industry,
- Cost-effective Vessel Washing (GG120), gives guidance on ways to reduce water and product loss by employing better cleaning practices,
- Increasing Product Output in Batch Chemical Manufacture (GG216), gives guidance on optimising the overall efficiency of batch manufacturing operations,
- Improving the Performance of Effluent Treatment Plant (GG175), helps companies to consider their effluent systems as an integral part of their plant operations which can lead to reduced waste and treatment costs,
- 'Effluent on line' is a web site that has been established by the programme to specifically help chemical and other process industry companies to spread information and techniques on good effluent management practices.[1]

The programme also has a range of material to help companies considering investment in clean technology solutions, including:

- Choosing Cost-effective Pollution Control (GG109),
- Cost-effective Membrane and Separation Technologies for Minimising Wastes and Effluents (GG37 & GG54),
- Investing to Increase Profits and Reduce Wastes (GG82),
- Life-cycle Assessment – an Introduction for Industry (ET257).

Access to the full range of programme information and services – including site visits for SME's (Small and Medium-sized Enterprises) to help them identify waste minimisation opportunities – is available.[2]

6 CONCLUSION

Chemical companies can make significant environmental improvements in their products, processes and management by the adoption of waste minimisation and clean technology measures.

Large-scale product and process enhancements can often seem to offer the largest improvement benefits. Whilst this might be true, it is almost certain that significant benefits can also be achieved from easier and less risky approaches using basic waste minimisation approaches. Adopting a combined step by step approach can ensure continual improvement, early payback on investments and greatest certainty of attaining long term improvement goals.

The author is grateful for the support of the Department of Trade and Industry and the Department of the Environment, Transport and the Regions who jointly fund the Environmental Technology Best Practice Programme. The support of many other Programme team members and the companies who have participated in the Programme is also acknowledged.

References

1. The web site address is *www.effluentonline.co.uk.*
2. The Environment and Energy Helpline is on 0800 585794 and the programme web site is at *www.etbpp.gov.uk.*

HOW GREEN IS MY PROCESS? A PRACTICAL GUIDE TO GREEN METRICS

Paul Smith

SmithKline Beecham Pharmaceuticals
New Frontiers Science Park
Third Avenue, Harlow
Essex   CM19 5AW

1 INTRODUCTION

The interest in Green Chemistry/Technology is growing rapidly and this paper is designed to provide some answers to the following questions:

1. What is Green Chemistry/Technology?
2. Why is it important?
3. How do we measure our progress from an early stage in development?

In order to address question 3, a review of published suggestions for measuring the "greenness" of reactions or processes is given. This is followed by the description of a template containing several parameters which builds on current methods and can provide a very useful tool in early development work. Examples are then provided to illustrate how the template can be used.

2 WHAT IS GREEN CHEMISTRY/TECHNOLOGY?

Many definitions of Green Chemistry/Technology[1] have been put forward but the following two statements provide a good general starting point:

- The discovery and application of new chemistry/technology leading to prevention/reduction of environmental, health and safety impacts at source

- Making best use of existing chemistry/technology to ensure minimum impact

More specifically, an ideal synthesis has been defined by Paul Wender[2] as one where the target molecule is:
- Prepared from readily available starting materials
- Prepared in one simple step
- Prepared by a safe process
- Prepared by an environmentally acceptable process
- Resource effective
- Synthesised quickly
- Synthesised in quantitative yield

The concept of atom economy, introduced by Barry Trost,[3] is also important and involves maximising the number of atoms of all raw materials that end up in the product, e.g. $CH_3P(Ph)_3Br$ in a methylenation reaction transfers a mass of only 14 out of 365 to the product. Taking this concept into account led Trost to define an ideal chemical reaction as one which is not only selective but is also just a simple addition in which any other reactant is required only in catalytic amounts.

The achievement of Green processes as described above obviously represents a major challenge and this is particularly true in the pharmaceutical industry where the chemistry effort is directed towards complex molecular targets with a need for early and rapid route definition.

2.1 Why is Green Chemistry Important?

For chemical companies, the pursuit of Green Chemistry/Technology principles can be expected to lead to a competitive advantage by delivering reduced costs/risks and by providing greater manufacturing flexibility. As legislation increases (e.g. Green House Gas taxes, limits on emissions) it makes very good sense to be adopting a pro-active approach in order to meet the challenges ahead. Also the achievement of greener processes can hopefully help provide a much needed improvement in the public image of the chemical industry.

The target is therefore to integrate Green Chemistry/Technology principles into synthetic and process design early in development and this leads on to the important question of how do we make decisions on which chemistries/processes are greener for a particular target molecule.

3 HOW DO WE MEASURE OUR PROGRESS?

In order to gain a complete understanding of the environmental impact of a product, a full life cycle analysis is needed (see Figure 1). This takes account of raw material usage, energy consumption and waste production not just for the process at the site of manufacturing but also for any raw materials and equipment production at other sites. In addition, the environmental impact of the product is also measured right through to ultimate disposal. Whilst this level of information would be extremely useful to have at an early stage it is just not practical at present.

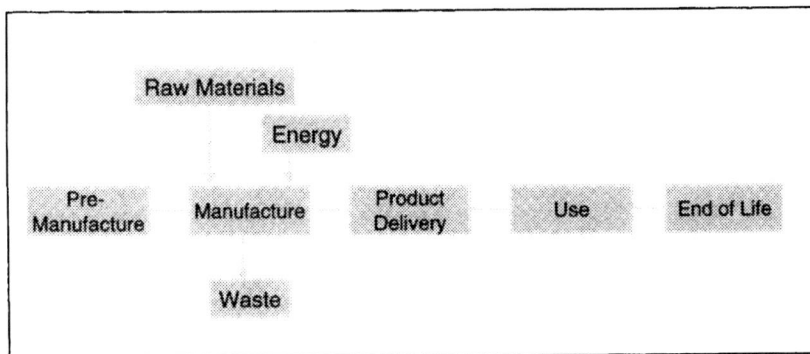

Figure 1 *Full life cycle analysis*

Some early metrics are therefore needed to:
- Allow progress to be monitored
- Provide the basis for sound discussion
- Highlight areas for improvement
- Provide evidence to Regulatory Authorities of the efforts made to minimise environmental impact

The big challenge here is to achieve simplicity in the metrics and yet retain the utility described above.

Several 'green' metrics have already been proposed and published over the years with perhaps the most well known being the E-Factor described by Roger Sheldon[4] as the 'Amount of waste produced per kg of product'. Table 1 shows some typical figures for E from the Sheldon publication. A further refinement suggested in this paper is to use the 'Environmental Quotient' which is the product of the E-Factor and Q, where Q is defined as the 'unfriendliness quotient' (e.g. NaCl, Q=1; heavy metal salts, Q=100).

More recently, Tomas Hudlicky[5] made the point that the traditionally used metrics of percentage yield and number of steps in a synthesis are too limited and he suggested the use of 'Effective Mass Yield' (EMY). This is defined as:

$$\frac{\text{Mass of desired product}}{\text{Mass of all non-benign material used}} \times 100$$

This is effectively the reciprocal of E quoted as a percentage but with materials like water, ethanol etc. omitted.

A further metric, suggested by Sheldon[4] for individual reactions, is 'Atom Utilisation' which is defined as:

$$\frac{\text{Molecular Weight of desired product}}{\substack{\text{Sum of molecular weights of all substances produced} \\ \text{in the stoichiometric equation for the reaction}}}$$

In summary, a wide variety of metrics have been quoted in the literature but it is clear that a single figure cannot provide a measure of 'greenness'. A suggested solution to this challenge is to use a template, containing a list of 'green parameters' which are a mixture of qualitative and quantitative assessments for a particular process (Table 2). These

Table 1 *Examples of E Factors*

	Product Tonnage	*E Factor*
Bulk Chemicals	10^4-10^6	<1 - 5
Fine Chemical Industry	10^2-10^4	5 - >50
Pharmaceutical Industry	10-10^3	25 - >100

metrics are relatively straightforward to obtain and can be usefully applied when comparing and/or improving routes.

Table 2 *Green Parameters*

Parameters	Purpose
Compound Number	Identification
Route Designation	"
Date of Assessment and Reference	"
Number of Chemistry Steps[a]	Complexity
Number of Purification Steps[b]	"
Number of Stages[c]	"
Number of isolated intermediates	"
% Overall Yield	Efficiency
List of Solvents used	Environmental Impact
List of extreme reaction conditions: (temps >130 and <-15°C, pressures >50 psi, dilution >20% volume/kg of product)	"
Key materials leading to known environmental/safety/health problems	"
Overall kg Solvent/kg Final Product	"
Overall kg Water/kg Final Product	"
Overall kg Input Material (excluding solvent and water)/kg Final Product	"
Total Waste/kg of Final Product (sum of 3 boxes above -1)	"
Overall kg Input Material (excluding solvent and water)/kg Final Product if all Stage Yields are 100%	A measure of 'atom economy'
Projected Peak Year Tonnage	Impact level
Catalytic Chemistry used	Highlights Greener methods
Asymmetric Chemistry used	"
Additional Comments	Space to elaborate on any findings

Notes to Table 2:

a) Chemistry Step

This is defined as a reaction which effects a structural change and gives an isolable product, but which may or may not be isolated in practice. (This includes salt formations where isolated, but not transient formations in acid/base extractions). When making route comparisons, a higher number of steps would imply that more complex chemistry is being used to reach the target.

> Example: Reduction of RNO_2 to RNH_2 would be a single step, even though it can be argued that the reaction goes through more than one transformation, i.e. through a nitroso and a hydroxylamine intermediate.

b) Purification Step

This is a step carried out to improve the purity or form of an intermediate or final product, and includes distillations, recrystallisations, chromatography and resolutions (by chromatography or formation of diastereomeric salts, isolation and neutralisation).

c) Stage

This is defined as a series of operations comprising one or more chemistry and / or purification steps followed by an isolation procedure to give the desired intermediate or final product as a solid, oil or in solution.

> Example of the difference between a stage and a step:

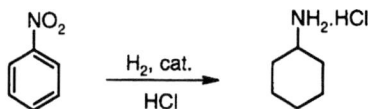

This may be achieved in one stage, but has three chemistry steps.
1. Reduction of nitro group
2. Reduction of aromatic ring
3. Hydrochloride salt formation

4 EXAMPLES

In order to illustrate the use of the template in route comparisons, an example is provided using a SmithKline Beecham compound, 32872. The flow diagrams are given for the route used for early supplies (Route A, Figure 2) and a 'new' route (Route B, Figure 3).

It is not appropriate to describe in detail the chemistry involved in the two routes but a comparison of the template results is shown in Table 3.

It can be readily seen that Route B offers significant advantages over Route A with fewer isolated intermediates and much lower solvent, water and input material requirements. The overall kg input material/kg of final product if all stages are 100% also show Route B to be inherently more efficient or more atom economic. The relative merits of using lithium aluminium hydride (LAH) or acrolein do of course need separate discussion.

A useful way to illustrate some of the data obtained is by way of bar charts and the next example (Figure 4) shows waste reduction as development work was carried out on 'compound A', where Route A is the original Medicinal Chemistry route and Route B is the route destined to become the manufacturing route.

Finally, Table 4 shows the key quantitative figures for four drug substances which are either in or very close to production.

These results are helpful in providing some benchmark figures which, as new green chemistry/technology options increase in the future, can be expected to be improved on for the compounds that are at the beginning of their development programme.

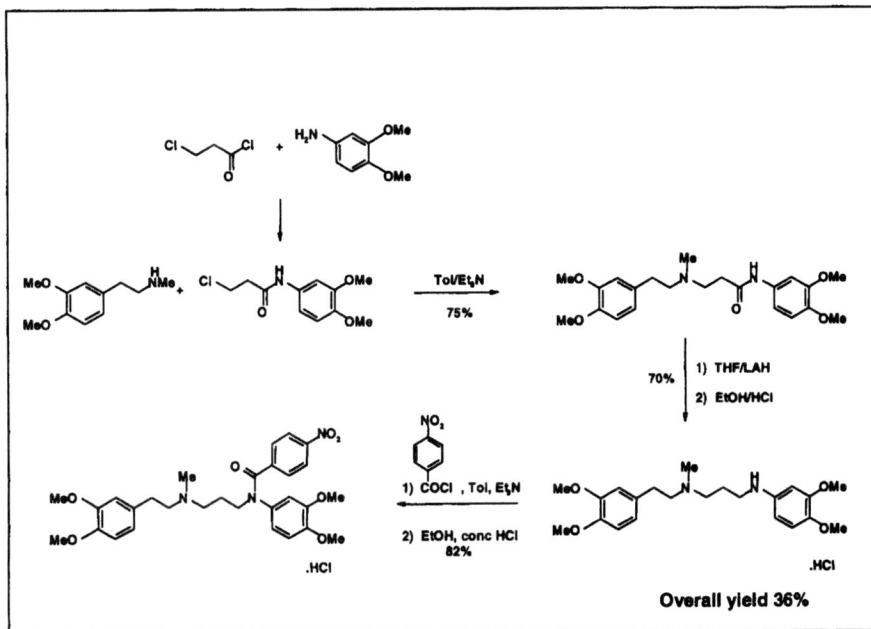

Figure 2 *Supply Route A to 32872*

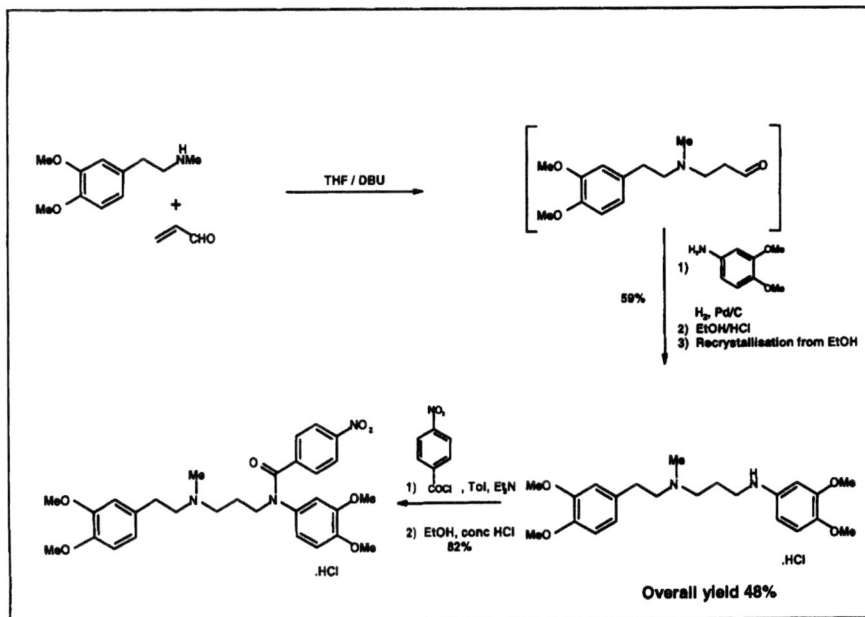

Figure 3 *Supply Route B to 32872*

Table 3 *Metric Examples - 32872*

	Route A	Route B
Number of Chemistry Steps	6	5
Number of Purification Steps	0	1
Number of Stages	6	5
Number of Isolated Intermediates	5	3
% Overall Yield	36	48
List of Solvents used	THF, Tol, MeOH, EtOH	THF, Tol, EtOH
List of extreme reaction conditions: (temps >130 and <-15°C, pressures >50 psi, dilution >20% volume/kg of product)	None	None
Key materials leading to known environmental/safety/health problems	LAH	Acrolein
Catalytic (non-stoichiometric) chemistry used	None	Hydrogenation
Asymmetric chemistry used	Not applicable	Not applicable
Overall kg solvent/kg of final product	48	33
Overall kg water/kg of final product	33	11
Overall kg input material (excluding solvent and water)/kg of final product	4.9	2.8
Total waste/kg of final product (sum of 3 boxes above -1)	85	46
Overall kg input material (excluding solvent and water)/kg of final product if all stages are 100%	2.5	1.7

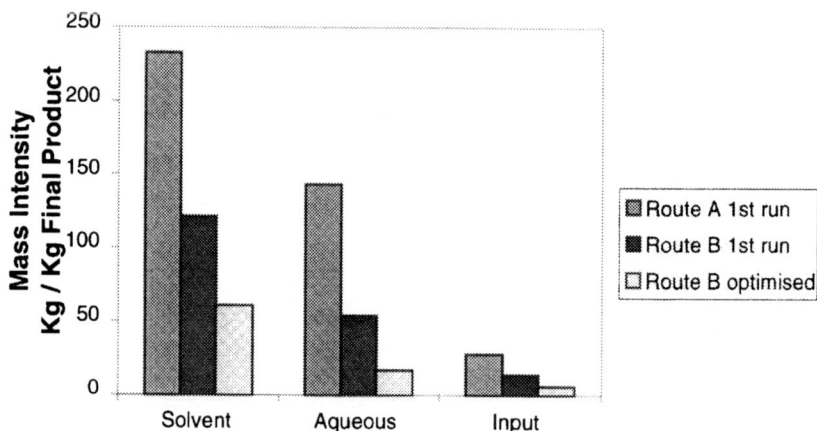

Figure 4 *SB Compound 'A' total wastes*

Table 4 Key Quantitative Figures for Four Drug Substances

	Compound A *Route B:* *5 Stages*	*Compound B* *Route C:* *4 Stages*	*Compound C* *Route B:* *3 Stages*	*Compound D* *Route C:* *4 Stages*
Kg Solvent	61	73	45	58
Kg Water	17	72	64	40
Kg Input Material	6	13	10	13.5
Kg Input Material if 100% yields	4	8.5	6.5	7
TOTAL WASTE	**83**	**157**	**118**	**110.5**

5 CONCLUSION

The search for metrics which can help measure progress in Green Chemistry is complex and potentially very time consuming. For example, even the simplest of metrics, % yield, can be misleading unless it is carefully defined (e.g. in-pot yield, isolated yield, yield after allowing for recovery of starting material etc.) It is however important not to be daunted by the complexity and to try to establish some metrics at an early stage in drug development. The template, incorporating both quantitative and qualitative data, can provide a practical first step towards a greater understanding of green processes.

Acknowledgements

My thanks go to the following people in SmithKline Beecham for the many discussions which led to the production of the template.

UK: Alan Curzons, Graham Geen, John Hayler, John Kitteringham, John
          Richardson, Luisa Freitas dos Santos
US: David Constable, Bob Hannah, Mike Mcguire, Lee Webb, Marvin Yu
     In addition I would also like to acknowledge the fine work done by Tom Ramsay, Robin Attrill, Richard Freer and Graham Slater on the synthesis of 32872.

References

1. A good detailed discussion on Green chemistry including 'the 12 principles of green chemistry' is provided in *Green Chemistry: Theory and Practice*, P. T. Anastas and J.C. Warner, Oxford University Press, 1998.
2. P. A. Wender, and B. L. Miller, *Org. Synthesis: Theory and Applications*, Edited by T. Hudlicky, 1993.
3. B. M. Trost, *Angew Chem. Int. Ed.*, 1995, **34**, 259; B. M. Trost, *Science*, 1991, **254**, 1471.
4. R. A. Sheldon, *Chem. Ind.*, (London) 1997, 12.
5. T. Hudlicky, D. A. Frey, L. Koroniak, C. D. Claeboe and L. E. Brammer, *Green Chem.*, 1999, 57.

PROCESS INTENSIFICATION: POTENTIAL IMPACT ON THE CHEMICAL INDUSTRY

C. Ramshaw

Centre for Process Intensification & Innovation
University of Newcastle upon Tyne
Newcastle upon Tyne NE1 7RU

1 INTRODUCTION

When the concept of process intensification was developed within ICI in the late 1970's, the original intention was to make big reductions in the cost of processing systems, without impairing their production rate. The term "Process Intensification" (PI) was used to describe the strategy of making dramatic (100 - 1000 fold) reductions in plant volume in order to meet a given production objective. It is well known that the cost of the main plant items (e.g. reactor, separators etc.) only represents around 20% of the cost of a production system, with the remainder being incurred by pipework, structural support, installation and so on. A major reduction in equipment size, with hopefully a high degree of telescoping of plant function, should lead to large cost savings by eliminating support structure, column foundations and long pipe runs.

The degree of miniaturisation involved is that needed to generate the cash savings required. Thus volume reductions of the order of 100 times must be our target in order to secure the desired impact on costs. While an individual intensified unit may cost a little more than the conventional equivalent, (although hopefully it will not) it must generate substantial overall savings in the cost of the process system. In addition, the process intensification philosophy should be applied across the whole spectrum of unit operations used in the plant. Bearing in mind the dramatic size reduction which is sought, process intensification will probably involve novel and unusual approaches to equipment design. It is not a strategy for the faint-hearted. Herein lies one of its main disadvantages, namely the lack of new design codes to engender confidence in those who specify new equipment. Radical and unconventional approaches will be the order of the day. Many searching questions will be posed, such as the need for turbulent flow in pipes, the use of batch rather than continuous operation and the application of merely terrestrial acceleration to multi-phase systems, to name but a few. It is a sobering thought that if chemical engineers were given a free hand to design the human digestive and metabolic system, our bodies would be much larger and require many kilowatts to operate them. On the other hand, nature operates unobtrusively with laminar flow in high density matrices (kidneys and lungs) on a semi-continuous basis, and copes with fouling problems by coughing. As scientists and engineers we should not be too arrogant to learn a few lessons from the natural world.

2 ADVANTAGES

While cost reduction was the original target for PI, it quickly became apparent that there were other important benefits, particularly in respect of improved intrinsic safety, reduced environmental impact and energy consumption. More recently it has become apparent that the whole business process for manufacturing chemicals could be revolutionised by an effective implementation of a PI strategy. These features are considered in more detail below and together they present an overwhelming incentive for the further development of intensified technology for the chemical and process industries.

2.1 Safety

Given the anticipated plant volume reductions, the toxic and flammable inventories of intensified plant are correspondingly reduced, thereby making a major contribution to intrinsic safety. This point has been well made by Trevor Kletz who has commented that "what you don't have can't leak".[1] Had intensified plant been available at the time, the Flixborough and Bhopal disasters would probably not have occurred.

2.2 Environment

With regard to the environment, the intensified plant of the future will be much less obtrusive, with the distillation and absorption towers of our present chemical complexes being replaced by more compact and inconspicuous equipment, which may be hidden by the boundary tree line. In addition, the cost of effluent treatment systems will be less, allowing tighter emission standards to be reached economically. The economic incentive to produce commodity chemicals in large centralised plant complex is likely to diminish or disappear with the application of PI. Thus distributed manufacture in smaller plant will be both feasible and economic, thereby avoiding the need to distribute hazardous material on the public transport system.

However, the most telling environmental influence of process intensification could well be in the development of new reactor design for truly clean technology. Rather than accept mere "end of pipe" solution, we must create fluid dynamic environments which allow the intrinsic chemical kinetics free rein. We then have a far better prospect of designing reactors which operate intensively and which give high selectivity. This would facilitate the delivery of a high quality product without an expensive downstream purification sequence.

2.3 Energy

The high heat and mass transfer coefficients which can be generated in intensified equipment can be exploited to reduce the concentration/temperature driving forces needed to operate energy transformers such as heat pumps, furnaces, electrochemical cells etc. This enhances the equipment's thermodynamic reversibility and hence its energy efficiency. For example we have shown at Newcastle that the application of elevated acceleration fields to a simple chlorine cell can reduce its voltage by over 0.4V.[2] Similarly the Rotex absorption air conditioner[3] which will soon be entering field trials, demonstrates a very high performance while avoiding the arcton/chlorofluoro carbon working fluids used in vapour compression air conditioners. Instead a cocktail of alkali metal hydroxide solutions is employed. Therefore innovative applications of PI thinking can improve our capacity to meet the energy and global warming targets which were agreed at Kyoto.

2.4 The Business Process

As pointed out above, the business process is likely to be profoundly influenced by a fully implemented policy of PI. A chronic problem in the manufacture of commodity chemicals is the cyclical nature of the industry. This is induced by over enthusiastic simultaneous investment in large-scale upratings which are then followed by an oversupplied market. The alternative scenario which is rendered feasible and economic by PI, is to uprate the plant in smaller steps in order to follow the market more closely. This should avoid the "boom and bust" with today's technology.

An inevitable consequence of PI is that the plant residence times will be reduced drastically (probably from hours to seconds). This implies that grade changes can be implemented quickly to satisfy rapidly changing markets. It also facilitates a "just-in-time" production philosophy with a consequent reduction in stock levels and the corresponding working capital involved.

Perhaps most important of all, in the pharmaceutical and fine chemical industry, it is worth questioning the cultural obsession with batch production which has held sway for the last century or so. This stems from the moment that the process development chemist reaches for a flask or beaker to produce trial quantities of a new molecule. Thenceforward the new process is locked in to a series of lengthy scale-up stages to satisfy the regulatory authority. Having started the patent "clock" on the discovery of the molecule these delays are vastly expensive over the lifetime of a blockbuster drug. At Newcastle this has led us to advocate the continuous "desk top" fine chemical process in order to help the drug companies commercialise their inventions more quickly. A typical production rate of 500 tonnes/year corresponds to a continuous flow rate of about 25 ml/s so a desk top comprising a continuous intensified reactor in conjunction with other intensified separation/crystallisation/drying modules, could comfortably meet the market demand. Scale-up problems are avoided because the laboratory scale is the full scale. A production facility for (say) ten products would involve ten desk top units each modulated and controlled by its own computer. Being continuous, the equipment will not be subject to the labour costs or potential contamination problems associated with inter-batch cleaning.

3 EXAMPLE

In the first case, the reactions of interest are those which are intrinsically fast and exothermic, but which are currently limited by the poor heat and mass transfer for rates achievable in a stirred pot. Existing technology routinely entails substantial hazardous process inventories, possible reactor runaway and indifferent product selectivity. Fast response reactors open up the possibility of switching to more severe process conditions which would be prohibited in conventional reactors in view of the tendency to degrade the product. It may therefore be possible to exploit a virtuous circle:- short residence time – higher temperature – faster kinetics – smaller reactor – shorter residence time.

Of the many options for intensified reactor design, two of the more attractive are being studied at Newcastle. They involve microprocessor technology (fine channels in a multi functional matrix) and a rotating surface of revolution (spinning disc reactor). A key characteristic of the latter is an ability to stimulate intense heat/mass transfer between a highly sheared liquid film and the rotating disc over which it moves, or the adjacent gas phase. This allows rapid reactions which involve viscous liquids or large exotherms to be precisely controlled. Following extensive discussions with potential industrial partners in the pharmaceuticals/fine chemicals area, it has become evident that there is considerable interest in the opportunities presented by spinning disc reactor (SDR) technology. A reactor design based on spinning discs provides an excellent heat and mass transfer environment for the reacting liquid and promises to overcome these disadvantages.

SDR technology offers the possibility of a step change in manufacturing operations, particularly with respect to the following attributes:-

1) Ability to cope with very fast exothermic reactions (corresponding to heat fluxes of upto 100 kW/m^2.
2) Low inventory/intrinsic safety (liquid film thickness are 50 - 200 μ).
3) Rapid response (liquid residence times are 1 - 5 seconds).
4) Easy cleaning.
5) Close control (due to short residence times).

However in general it is perceived that the biggest obstacle to the adoption of SDR technology will be cultural issues rather than technology. In particular, chemists involved in process development, have both a lack of awareness of SDR and a fear of "mechanical" innovations.

At Newcastle we are tackling this problem by manufacturing prototype SDR's in our workshops and then arranging to have them operated in the laboratories of our industrial collaborators. Initial results are very promising and it is anticipated that several joint projects will emerge in order to perfect the technology for each client's application. Ultimately it is the intention to have simple proven versions of SDR's available when the process route is being developed by the chemist. Hopefully this will encourage the adoption of a continuous processing strategy from the outset, because once beakers or flasks are used in the initial process development it is very difficult thereafter to gain support for a continuous option.

It is recognised that a typical fine chemical/drug process involves many operations in addition to the reaction stage. These may be extraction, precipitation, solids removal, drying, distillation etc. In order to bring the desk top plant to reality, intensified versions of the relevant conventional equipment must be made readily available to the process research chemist, otherwise we will end up with the same old pots and pans as before. Although this is a challenging target, the business benefits justify its enthusiastic acceptance.

4 CONCLUSIONS

1) A strategy of process intensification requires a step change in the philosophy of plant and process design.
2) If effectively implemented it will revolutionise the business process and lead to major improvements in environmental acceptability, energy efficiency, intrinsic safety and capital cost.

3) A major cultural change is required on behalf of chemists, engineers and managers and it is this rather than technical difficulty, which represents the main obstacle to progress.

References

1. T. Kletz, 'What you don't have, can't leak', *Chem & Ind*, 1978, May 6.
2. H. Cheng, K. Scott, and C. Ramshaw, 'Electrochemistry in a Centrifugal Field', Proc. Conf. on Process Innovation and Intensification, Manchester UK (Inst. Chem. Eng.), 1998, Oct 21-22.
3. C. Ramshaw, and T. L. Winnington, 'An Intensified Absorption Heat Pump', *Proc. Int. Refrigeration*, 1998, **85**, 26.

PROCESS INTENSIFICATION: CHOOSING THE RIGHT TOOLS

C. de Weerd

Akzo Nobel Chemicals Research Arnhem
P.O. Box 9300
6800 SB Arnhem
The Netherlands

1 SYNOPSIS

Process Intensification provides a new incentive in the assessment of chemical processes. In this report a gas-liquid reaction forms the basis of a comparison between conventional and novel reactor concepts. The considered reaction can be applied in all of the described reactor concepts, in some cases however a process modification is needed. For the application of Process Intensification a good insight is required for both the reaction as well as the tools (reactor systems) in which the reaction is being applied. Regarding the novel reactor concepts however design data are scarce, which induces uncertainties to come to a correct validation of the potentials.

2 NOTATION

a	G/L interfacial area per unit volume	$[m^2/m^3]$
d	nozzle diameter of ejector	$[m]$
d_{32}	Sauter mean bubble diameter	$[m]$
d_h	hydraulic diameter of mixing element channels	$[m]$
D_l	diffusion coefficient	$[m^2/s]$
k_l	liquid-side mass transfer coefficient	$[m/s]$
u_c	superficial velocity continuous phase in contactor	$[m/s]$
u_d	superficial velocity dispersed phase in contactor	$[m/s]$
V_{ej}	volume of ejector	$[m^3]$
We_c	critical Weber number at break-up (≈ 1.1)	$[-]$
ε_v	void fraction of mixing element	$[-]$
ε_c	continuous phase hold-up	$[-]$
ε_d	dispersed phase hold-up	$[-]$
v_c	kinematic viscosity continuous phase	$[m^2/s]$
ρ_c	density continuous phase	$[kg/m^3]$
ρ_d	density dispersed phase	$[kg/m^3]$
σ	G/L interfacial tension	$[N/m]$
$\in$	power dissipation per unit mass	$[W/kg]$

3 INTRODUCTION

In the past few years a novel approach to chemical engineering has emerged. This new way of treating processes is called Process Intensification (PI). Although many definitions exist for PI, the main goal is to achieve considerable size reductions in chemical plants. Consequently a smaller plant will lead to improvements in the field of HSE, energy consumption and waste reduction. Recently an overview of PI incentives has been given by Stankiewicz et al.[1]

Parallel to PI, a number of equipment vendors have been developing a number of interesting new apparatuses. This new equipment is mainly focussed on the reaction section in the chemical process. Although the size of plant often is determined by the downstream processing part, i.e. separating sections such as distillation and extraction, a first step in the PI approach might be the size reduction of the reactor.

In this report the merits of a number of these novel reactors will be compared to the conventional approach for a given gas-liquid reaction.

4 REACTION SYSTEM DEFINITION

For the comparison of a number of techniques a gas-liquid reaction is being considered given by the following equation:

$$A_g + B_l \longrightarrow C_g + D_l$$

Scheme 1 *Typical gas-liquid reaction*

Generally speaking this reaction can be classified as a chemical enhanced absorption. A model reaction for this system might be the chlorination of toluene. In this case the reaction can be described by:

$$Cl_2(g) + toluene(l) \longrightarrow HCl(g) + chlorotoluene(l)$$

Scheme 2 *Chlorination of toluene*

Let us consider some characteristics of this particular system. The heat of reaction for a chlorination is roughly -100 kJ/mole, corresponding with an adiabatic temperature rise of approximately 500 K. Therefore the system can be identified as being moderately exothermal. Heat removal will play an essential role in the design of a suitable reactor.

In the model system two important phenomena take place. First, of course, there is the intrinsic kinetics of the reaction between components A and B. Secondly there is the mass transfer of the gaseous compound into the liquid. When dealing with relatively fast reactions, which is the case for chlorinations, the overall reaction rate might be mass transfer controlled. It must be noted that this limitation should always be checked when dealing with such a process. Obviously it requires a lot of know how of the considered system. Viewed in that light, it is interesting to note that in some cases such a system nevertheless might be kinetically controlled.[2]

Another aspect is the gas-liquid ratio in such a process. By applying the physical properties and the ideal gas law, it can easily be calculated that the G/L ratio is approximately 250 at 1 bar. Obviously this ratio is quite large and also is of great influence in the choice of the reactor.

Summarising, we have to deal with potential limitations in the field of heat transfer, mass transfer, and fluid mechanics.

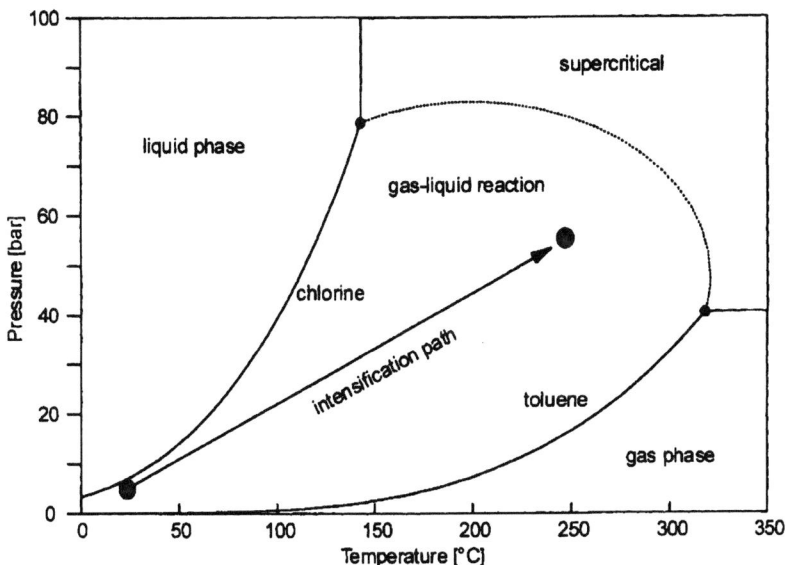

Figure 1 *Reaction contacting regimes*

For the intensification of this particular process, two process parameters play an important role, i.e. temperature and pressure. By increasing temperature the intrinsic reaction rate will increase due to the rise of the reaction rate constant. By increasing pressure, the solubility of the gaseous reactant will increase resulting in a higher concentration in the liquid. A pressure rise however is not always recommended when dealing with potential explosive components. For safety reasons, reactors for this type of components have to be designed for high pressures to absorb any possible explosions.

In Figure 1 the pressure-temperature diagram of the toluene-chlorine system is given. Several contact regimes can be distinguished. At low temperatures the reaction system is a homogeneous liquid. At high temperatures and low pressures the reaction takes place in the gaseous phase. When both temperature and pressure are high, the system is in the supercritical area. The remaining area in Figure 1 deals with the considered gas-liquid contacting regime.

In the case of intensifying this model reaction, taking place at 25°C and 5 bar, an obvious step would be to increase both temperature and pressure. This is shown by the arrow which is drawn in the figure. At this point however only the intrinsic kinetics are being increased. Due to this improvement the overall reaction will probably become mass-transfer limited, also a limited heat transfer might give rise to limitations. Therefore in the intensified working point it is necessary to use enhanced reactor equipment to overcome these hurdles. An elaborate summary describing the performances of a number of applicable gas-liquid contacting devices has been provided by Lee et al.[3]

5 CONVENTIONAL APPROACH

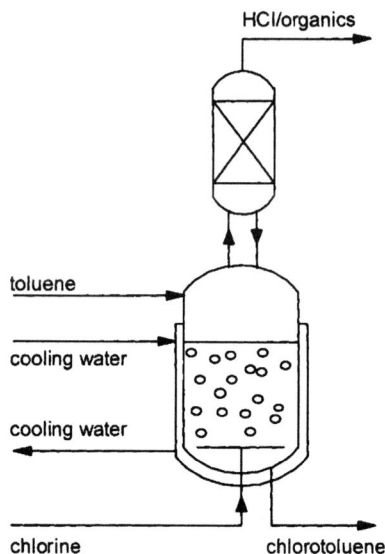

Figure 2 *Conventional gas-liquid contactor*

A typical conventional reactor to use for the model system is depicted in Figure 2. This continuous reactor can be characterised as being a continuous stirred tank reactor (CSTR) operating at cross-flow conditions. The liquid reactant is fed at the top of the reactor. The gaseous reactant is fed at the bottom via a gas sparger. Part of the reaction heat is removed by using a jacketed cooler. Optionally use can be made of either an internal or external heat exchanger. However the majority of this heat is removed by applying an external condenser to condense the vaporised organic components. Also the liquid reaction mixture acts as a kind of heat sink by absorbing part of the reaction heat.

Although the gas-liquid ratio of the reactants can be high, the gas hold-up is relatively small, e.g. 10%. In this way a sufficient gas-liquid contact area can be provided. A k_l a value of 0.01 s^{-1} is typical for this conventional reactor.[3] Another advantage is the improved heat transfer to the jacketed cooler, because the heat transfer coefficient of a liquid is considerably higher compared to a gas.

6 BUSS LOOP REACTOR

A good example of an intensified reactor for the model system is the Buss loop reactor concept as being offered by Kvaerner Process Technology (formerly Buss AG). This reactor is depicted in Figure 3. Reactor volumes range from 20 litres to 100 m^3 in practice.[3]

The gaseous reactant is fed at the head space of the autoclave, where it is drawn by the liquid reaction mixture through a mixing nozzle. In this way a well-dispersed mixture is created in the venturi mixer. Circulation takes place by using a special pump which can handle high gas/liquid ratios and even suspended catalysts. An external heat exchanger

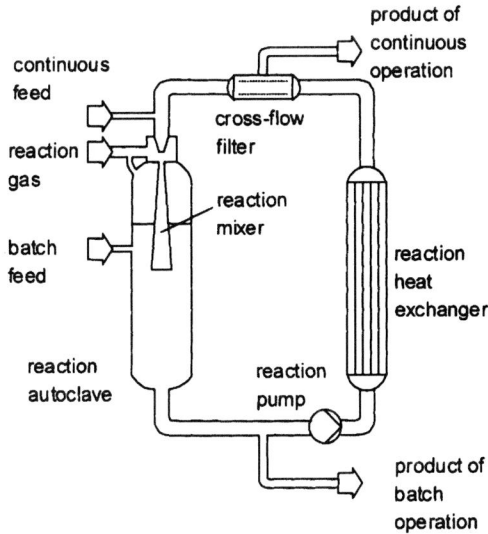

Figure 3 *Buss loop reactor*

allows for sufficient heat transfer of the reaction system. In order to retain catalyst within the system, use can be made of a cross flow filter.

The loop reactor concept can be regarded as being proven technology, Kvaerner claims e.g. that more than 200 units have been sold in the field of oleochemicals production. Also a considerable number of publications have been dedicated to this particular reactor concept.[4-12]

In addition some remarks can be made on the operating conditions of the loop reactor. Basically the loop reactor is a device especially suitable for batch reactions e.g. hydrogenations. In this type of reaction a gaseous compound is being consumed and no gaseous by-product is being produced. Operating in a continuous mode will demonstrate the disadvantage of the CSTR, which is incomplete conversion from a theoretical point of view. Another important aspect is the fact that in continuous mode the recycle ratio, i.e. the volumetric ratio of recycled volume and liquid feed, must be greater than 20, to guarantee sufficient mixing.[4] At the same time the gas-liquid ratio in the venturi should be smaller than 3 for proper operating, according to literature data.[9] By using high gas-liquid ratios as mentioned in Section 4, this means that a large amount of liquid must be recycled by the pump.

Cramers et al.[10] show that the mass transfer can be described by:

$$a = \frac{6\varepsilon_d}{d_{32}} \tag{1}$$

$$d_{32} = 0.62 \left(\frac{We_c}{2}\right)^{0.6} \left(\frac{\sigma^3}{\rho_c^2 \rho_d}\right)^{0.2} \left(\frac{1}{\varepsilon}\right)^{0.4} \left(\frac{1+\varepsilon_d}{1+0.2\varepsilon_d}\right)^{1.2} \tag{2}$$

$$k_l = 0.3\sqrt{D_l}\left(\frac{\varepsilon}{v_c}\right)^{0.25} \tag{3}$$

$$\in = \frac{\pi/_8 \, \rho_c \, d^2 \, (u_c + u_d)^3}{(\varepsilon_c \, \rho_c + \varepsilon_d \, \rho_d) V_{ej}} \tag{4}$$

The k_l a value ranges from 0.01 to 2 s⁻¹, and values of 2 to 5 kW/m³ are typical for the power dissipation.[3]

7 STATIC MIXER TECHNOLOGY

Another approach is to make use of static mixing devices for improved gas-liquid contacting. In contrast with the loop reactor, this device is a plug flow reactor, and therefore less backmixing occurs. The amount of literature data on static mixers is impressive. Applied design information can also be obtained by joining the High Intensity In-Line Mixing Research Consortium (HILINE) carried out by the British Hydromechanics Research Group in the UK.[13] Sulzer[14,15] and Kenics[16] mixers, especially, have found wide acceptance within the chemical industry. In most cases volumes are smaller than 10 m³ for static mixing devices.[3]

Limiting ourselves to Sulzer technology, two interesting mixer types emerge. The Sulzer Mixer Reactor (SMR) is ideal for exothermal reactions, but is, however, only applicable for viscous fluids, e.g. polymers. The Sulzer SMV mixer on the other hand can be applied for mixtures in which gas is the continuous phase. For this reason jacketed cooling is of no use because the heat transfer coefficient for a gas is rather limited. An important remark must be made on this point: only gas-liquid mixtures with a G/L ratio up to 10 can be handled with the SMV mixer.[14] Therefore, when dealing with the model system, solutions must be found to decrease this ratio within the reaction system.

An obvious measure would be the compression of the gas to meet this constraint. Because a static mixer basically is a plug flow device, this would probably lead to a steep temperature rise at the entrance of the reactor. This situation is depicted in Figure 4. In most cases this will not be acceptable. Therefore a viable option might be the recycle of the liquid stream in the system, which is shown in Figure 5. By recycling the liquid, compression of the gas can be minimised. Another advantage is a drastic reduction in adiabatic temperature rise, a 50% liquid recycle will halve this temperature rise. A disadvantage of this intervention is however that the effect of side reactions is more pronounced. This drawback is strengthened due to lack of adequate cooling.

Considerable data is available for static mixers. With the aid of the following relations applicable to Sulzer SMV mixers, a dedicated design can be made.

$$d_{32} = 0.21 d_h \left(\frac{\varepsilon_v}{u_c + u_d} \right) \left(\frac{\sigma}{\rho_c \, d_h} \right)^{0.5} \left(\frac{(u_c + u_d) d_h}{\varepsilon_v \, v_c} \right)^{0.15} \tag{5}$$

$$k_l = 0.0062 \frac{D_l}{d_h} \left(\frac{u_c \, d_h}{v_c \, \varepsilon_v} \right)^{1.22} \left(\frac{v_c}{D_l} \right)^{1/3} \tag{6}$$

$$\in = \frac{0.9 \rho_c \, (u_c + u_d)^3}{2 \varepsilon_v^3 \, d_h \, (\varepsilon_c \, \rho_c + \varepsilon_d \, \rho_d)} \tag{7}$$

The power input ranges from 10-700 kW/m^3, and k_l a values ranging from 0.1 to 3 s^{-1} are typical.[3]

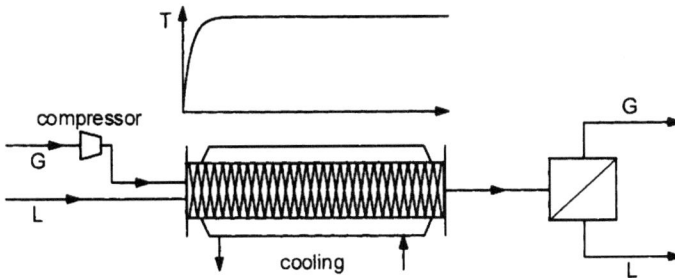

Figure 4 *Static mixer reactor concept without recycle*

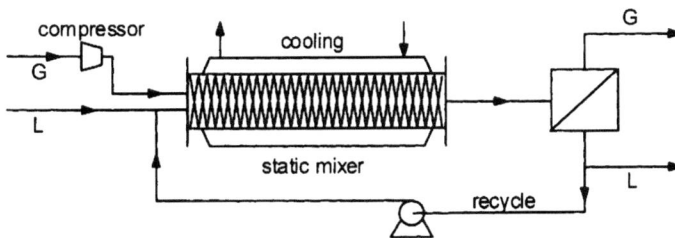

Figure 5 *Static mixer reactor concept with recycle*

8 MICROREACTOR TECHNOLOGY

An interesting development in the field of process intensification lies in the current availability of microreactors. These reactors can be constructed out of steel, and are made of thin slices with openings varying from 10 to 100 μm. Due to these small dimensions heat transfer coefficients can be reached of 20 kW/m^2K.[1]

Because microreactors are just getting off the age of childhood there is only limited information available for setting up a process design for a given reaction system. Nevertheless the results of microreactor technology for a number of actual processes have been published.[17] In the model system considered here, advantage can be taken of the enhanced heat exchanging capacity. For instance a combination of a microreactor together with a plug flow reactor (static mixer) might be an attractive set-up. The microreactor could be used for reaching a conversion up to e.g. 90% as well as removing the corresponding amount of reaction heat. The plug flow reactor could serve for finishing off the reaction to a complete conversion. The set-up is shown schematically in Figure 6. For this reactor combination it is assumed that the gaseous reactant has to be compressed in order to obtain a reasonable gas-liquid ratio. Probably this ratio is less critical for a microreactor relatively to a static mixer.

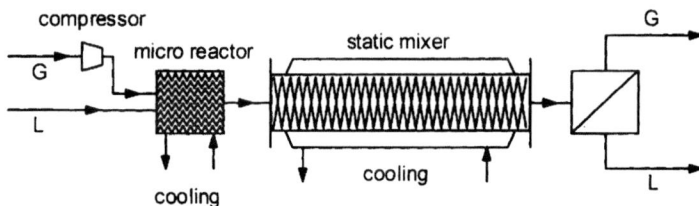

Figure 6 *Microreactor in combination with a plug flow reactor*

9 ROTATING PACKED BED

In order to improve the heat transfer in a reactor, use can be made of gravitational forces. This concept is used in the spinning disc reactor (SDR) as developed at Newcastle University. The reaction mixture flows in a thin layer in axial direction over a rotating disc. A typical heat transfer coefficient is 10 kW/m^{2}K. This reactor however is dedicated for liquid-liquid reactions. Especially condensation reactions can be enhanced by removing the gaseous by-products thus shifting the chemical equilibrium to the right.

A modification of this device is the so-called rotating packed bed reactor (RPB), as depicted in Figure 7. In this case a packing is attached to the rotating disc. In contrast with the SDR, this reactor is capable of handling gas-liquid reaction mixtures. The liquid reactant can be fed in the centre of the disc, the gaseous reactant moves countercurrently in the packing. Details on this particular reactor are provided by Burns et al.[18] Rotational speeds up to 1000 rpm are reported, corresponding with residence times in the range of 1s. The liquid flow rate lies around 1 m/s. The RPB reactor is of special interest because of its capability of treating reaction mixtures with relative high gas-liquid ratios. Burns mentions a G/L ratio of approximately 70, therefore especially this reactor seems suitable for the reaction model mentioned in this report.

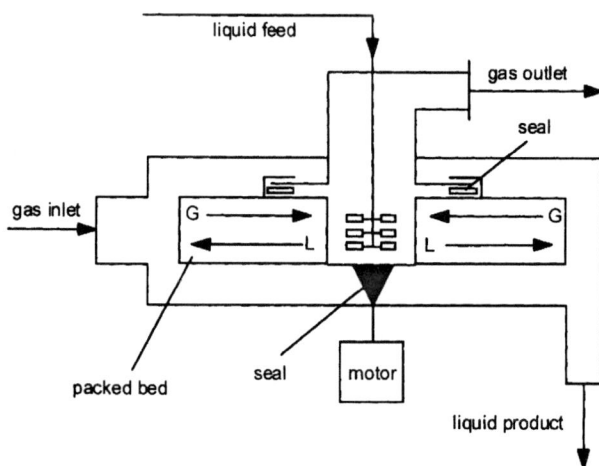

Figure 7 *Rotating packed bed reactor*

The RPB reactor has already been considered for a commercial process to produce hypochlorous acid. This component can be produced by reacting chlorine with aqueous caustic soda. The RPB set-up proved to be an attractive reactor in view of size reduction as well as in providing an improved product quality.[19]

10 CONCLUSIONS

Process Intensification offers an interesting incentive for the assessment of existing chemical processes. By comparing a number of established technologies as well as novel reactor set-ups, some distinctions can be made.

In contrast with conventional reactor concepts, rules for the design of novel reactors are not always present. In order to come to a valuable assessment of these new reactors, additional information should become available to investigate the full potential. This information should concern e.g. data on mass/heat transfer characteristics and fluid mechanics.

Considering gas-liquid reactions the high gas-liquid ratio is an important reason for rejecting a reactor in some cases. By modifying the process however the reactor nevertheless might be suitable.

In view of Process Intensification it seems worthwhile to perform a benchmark test by using a model system. It should be noted that detailed information of this system should be available regarding the intrinsic kinetics as well as the vapour-liquid equilibria. The performance as well as the PI potentials of a novel reactor can thus be studied with the aid of these data.

Summarising it might be concluded that Process Intensification demands an appropriate knowledge of the chemical process as well as a good understanding of the tools (i.e. reactors) to handle the process.

References

1. A. I. Stankiewicz and J. A. Moulijn, *Chem. Eng. Prog.*, 2000, **96** (1), 22.
2. G. F. Froment and K. B. Bischoff, *Chemical Reactor Analysis and Design*, 1st Edn., Wiley, Chichester, 1979, pp. 318-320.
3. S.-Y. Lee and Y. Pang Tsui, *Chem. Eng. Prog.*, 1999, **95** (7), 23.
4. G. M. Leuteritz, P. Reimann and P. Vergéres, *Hydrocarbon Process.*, 1976, **55** (6), 99.
5. L. L. van Dierendonck, G. W. Meindersma and G. M. Leuteritz, *Proc. Eur. Conf. Mixing*, 1988, 6th, 287.
6. C. A. M. C. Dirix and K. van der Wiele, *Chem. Eng. Sci.*, 1990, **45** (8), 2333.
7. P. H. M. R. Cramers, L. L. van Dierendonck and A. A. C. M. Beenackers, *Chem. Eng. Sci.*, 1992, **47** (9-11), 2251.
8. P. H. M. R. Cramers, A. A. C. M. Beenackers and L. L. van Dierendonck, *Chem. Eng. Sci.*, 1992, **47** (13/14), 3557.
9. P. H. M. R. Cramers, A. A. C. M. Beenackers and L. L. van Dierendonck, *Polytech. Tijdschr.: Procestech.*, 1993, **48** (2), 42.
10. P. H. M. R. Cramers, L. Smit, G. M. Leuteritz, L. L. van Dierendonck and A. A. C. M. Beenackers, *Chem. Eng. J.*, 1993, **53** (1), 67.

11. P. H. M. R. Cramers, R. F. Duveen and L. L van Dierendonck, "Process Intensification with Buss loop reactors", *Proc. Eur. Symp. on Catalysis in Multiphase Reactors*, 1994.
12. D. Nardin and P. H. M. R. Cramers, *Spec. Chem.*, 1996, **16** (8), 308, 310, 312.
13. A. Green and M. Zhu, "High-Intensity In-Line Mixing and Mass Transfer, Future Work of HILINE Project 1997-2000", BHR Group Ltd, Cranfield, 1996.
14. F. Grosz-Röll, J. Bättig and F. Moser, *VT Verfahrenstechnik*, 1983, **17** (12), 698.
15. F. A. Streiff, *Proc. Eur. Conf. Mixing*, 1979, 3rd (Vol. 1), 171.
16. Chemineer Inc., Kenics Static Mixers, *KTEK Series*, May 1988, Chemineer, Dayton, OH, 32 pages.
17. W. Ehrfeld, V. Hessel and V. Haverkamp, in *Ullmann's Encyclopedia of Industrial Chemistry*, 6th Edn., Wiley-VCH, 1999, Chapter Microreactors.
18. J. R. Burns, J. N. Jamil and C. Ramshaw, *Chem. Eng. Sci.*, 2000, **55**, 2401.
19. D. Trent, D. Tirtowidjojo and G. Quardener, *BHR Group Conf. Ser. Publ.*, 1999, **38** (Process Intensification for the Chemical Industry), 217.

SULPHONES BY OXIDATION - THE DEVELOPMENT PERSPECTIVE

David A Jackson*, Howard Rawlinson & Oskar Barba

Zeneca
Process Technology Department
Huddersfield Works, Leeds Road
Huddersfield
Yorkshire HD2 1FF

1 INTRODUCTION

The purpose of this paper is to present a perspective of the "Green Chemistry" Issues typically pertinent to the manufacture of Fine Chemicals. This is exemplified by describing the process development of the oxidation of a thioether to a sulphone. By "Fine Chemicals" I am referring to products generally described as consumer products, pharmaceuticals, agrochemicals, electronic chemicals and similar speciality effect chemicals. The scale of manufacture of a product in this class can vary from tens of kilograms to thousands of tonnes per annum. That class is distinct from bulk chemicals manufactured in excess of hundreds of thousands of tonnes per annum.

1.1 Business Imperatives

A common representation of the overall business process for a Fine Chemical Organisation is described in Figure 1.

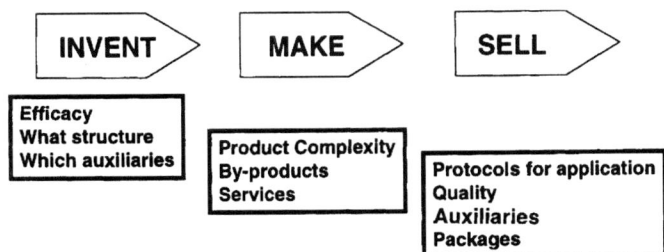

Figure 1 *Representation of a typical business process*

The environmental implications of a product manufacture are dependent upon the nature of the invented product, how it will be used, and for what period. The components listed in the boxes in Figure 1 have environmental implications. For example, the amount and nature of waste is related to manufacturing scale, efficacy and the structure of the product. The interdependency of the different components in the business process add to the complexity of managing the environmental implications of the manufacture. From this we draw the assertion that the best way to minimise the environmental impact of manufacture is to consider the issue holistically for the entire business process and to start thinking early in the lifetime of a project leading to a potential new product.

The environmentally relevant business drivers notable for the manufacture of fine chemicals are product life cycles and product registration and regulation. The product life cycles are typically 5-30 years, before a better product either supersedes the product or patent protection expires and competitive manufacture significantly reduces the profitability of the business. Consequently, there is strong benefit in minimising the time between invention and sales. To meet this objective, businesses often attempt to make the critical path for delivery of first sales dependant upon obtaining the regulatory approval for sale of the new product. This is represented in Figure 2. The data presented to the regulatory authorities when seeking registration must be derived from product with a composition typical of future manufacture. If manufactured product quality is significantly different, sales are not possible until re-registration takes place. The composition and quality of the product are clearly dependent upon the manufacturing technology. The two issues of minimising time to market and ensuring regulatory compliance place significant pressures to reduce time and process changes in the route selection and technology design phases of a project. This obviously constrains the amount of time available to evaluate and develop the manufacturing technology with the minimum environmental impact. Because of the limited product lifetimes there is commonly no second chance to develop 'better' technology at a later date.

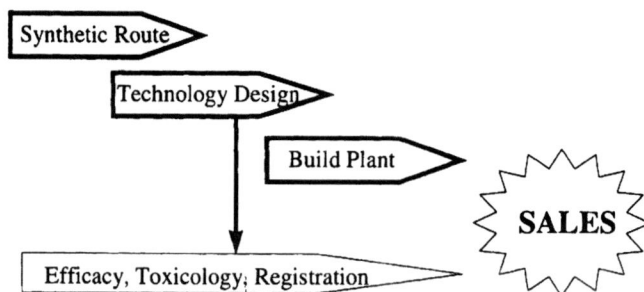

Figure 2 *Principal activities in new product introduction*

1.2 A Definition of Clean Technology

There are many definitions of clean technology, for practical purposes clean chemical manufacturing technology is the choice and organisation of synthesis route, reagents, engineering and by-product processing and disposal in order to minimise the overall environmental impact of the manufacture of the desired product. More succinctly, this can be described as, "minimise the environmental impact of the auxiliaries and by-products". This entails:

- atom efficient synthesis route
- maximising selectivity
- minimising process auxiliaries
- application of recovery and recycles
- by-product abatement and treatment

Chemical yields can be low provided the chemical selectivity is high and recovered starting materials or by-products can be recycled.

2 DEVELOPMENT OF A SULPHONE MANUFACTURING PROCESS

A process suitable for the conversion of thioether **1** into sulphone **2** was required at a manufacturing scale in excess of 1000 tonnes per annum. The initial research process entailed oxidation of **1** with hydrogen peroxide (3 mole/mole) in excess acetic acid. The variable yield of 60-70% was obtained over several batches during the manufacture of 100 kg of **2**.

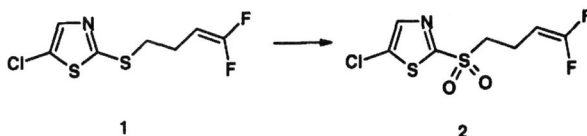

Our objectives of the development were to:

- A process with acceptable cost
    - Increase the yield
    - Best oxidant £/mole
    - Reduce process complexity (capital)
    - Improve the volume productivity
- Minimise the environmental impact
    - Increase the yield
    - Reduce the acetic acid recovery/recycle/disposal
    - Avoid toxic by-products

Typical oxidants for the conversion of a thioether to a sulphone are listed in Table 1.

The use of hydrogen peroxide as an oxidant is generally desirable due to cost and ease of use. Most of the development objectives could be met by removing or replacing acetic acid from the process. One of the seminal reports of the oxidation of thioethers, with hydrogen peroxide was made by Freyermuth,[1] in which, sodium tungstate is used as a catalyst in the process.

As both **1** and **2** are liquids above 20°C we evaluated the oxidation of neat **1** with hydrogen peroxide. No reaction took place. We then evaluated the effect of catalysts in

Table 1 *Cost and Implications of Different Oxidants*

Alternative Reagents	Implications	Approx. £/kg mole
Persulphate	Solvent, sulphate waste	350
Perborate	Solvent, borate waste	100
Peracids	Acid recycle	>60
Chlorine/hypohalide	Inorganic waste	50
Hydrogen peroxide	Catalyst required	25
Oxygen/air	Auxiliary required	?

this two liquid phase system. There are many reports of the use of phase transfer catalysts to aid this type of transformation.[2]

A variety of metal salts, transfer agents, complexes and liquids were examined. Some gave higher yields and rates than others and acceptable yields. However, all of them suffered from either; no activity, off-line catalyst preparation, tar formation, product contaminated with catalyst or a requirement for a co-solvent. Emulsion and tar formations were troublesome. This resulted in yield loss and contaminated product after difficult work-up procedures that would be unsatisfactory for large-scale production. The best overall result was obtained using sodium molybdate (5 mole%) and tris[2-(2-methoxyethoxy)ethyl]amine 3 (TDA-1, 5 mole %) at 60°C. The yield was >90% and the catalyst system could readily be removed by water washing.

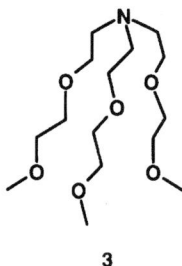

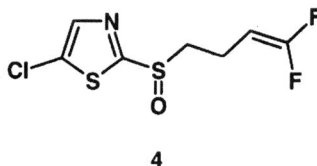

3   4

The anticipated large-scale costs and availability of TDA-1 prompted us to investigate the role of the catalyst system in order to allow us to reduce the molar ratio. We found that as the reaction progressed, the pH fell from the initial ambient value of pH 7.8 to <3 by the completion of the reaction. We attribute the pH change to formation of small quantities of sulphonic acid derivatives. The observation of pH change was puzzling in two ways; how did the pH drift effect the catalyst system and what was the optimum pH for the reaction.

It is known that molybdate anions undergo oligomerization. The species present are dependent on the pH of the solution. Csany and Jaky[3] have examined this phenomenon in the context of peroxomolybdate species. The phase-transferred oxidant is believed to be $MoHO^-Q^+$ in which Q^+ is a quarternary ammonium ion. The role of TDA-1 as an auxiliary in a phase transfer process of anions is generally envisaged as a ligand. The lone pains on the oxygen co-ordinate to metal cations, the resulting cation complex is more hypophilic than the hydrated metal cation and is able to pass into an organic phase as counterion to a suitable anion.[4] We ran the conversion of 1 to 2 at different pH values maintained throughout by the addition of sodium hydroxide solution. The reaction was sampled after 3 hours and the composition of the mixture was determined by gas chromatography. Three components were present; the thioether 1, sulphone 2 and sulphoxide 4. Figure 3 displays the relative ratios of the components (normalised to 100% total) in the reaction mass at the given pH after 3 hours. No other products were formed to any significant extent in these reactions.

The sulphoxide 4 is, as expected, an intermediate. The results indicate that the reaction proceeds fastest below pH5. At this pH and lower, TDA-1 is largely present in the protonated ammonium form, unavailable for complexation of sodium ions. We speculated that the transfer mechanism of TDA-1 in this reaction might be as an ammonium counter ion to peroxomolybdate. We evaluated this by testing three alternative

% Yield

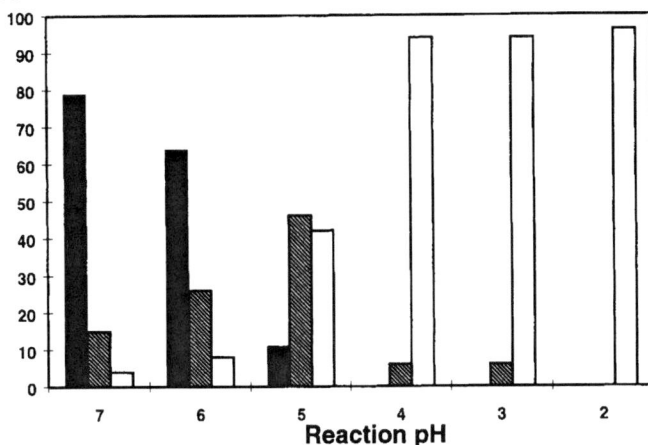

Figure 3 *Relative yields of 1, 3 and 4 from reactions at different pH.*
Thioether 1 solid bar, Sulphone 2 clear bar, Sulphoxide 4 hashed bar.

amines and poly(ethyleneglycol) in the reaction. The results are reported in Table 2.

No reaction took place with hydrogen peroxide and tri-n-butylamine in the absence of sodium molybdate. The observation that tri-n-butylamine is an effective catalyst led us to evaluate the reaction with a range of structurally different amines. The results are reported in Table 3.

All the thioethers are liquids at 60°C with the exception of dibenzothiophene. The reactions were performed neat, except dibenzothiophene was reacted in toluene. High yields are obtained. In the absence of sodium molybdenate and tri-n-butylamine the reactions were very slow. There was no evidence for tar formation in the catalysed reactions. The reactions were worked-up by washing with dilute acid. Phase separation was achieved easily. This effectively removed from the product the molybdate (<50 ppm by ion coupled plasma emission spectroscopy) and amine (<0.1% by gas chromatography).

Tri-n-butylamine is converted to the amine oxide[5] with hydrogen peroxide at 90°C, but the amine oxide was not detected (<1% by ^{1}H-NMR) during the reaction of **1**. The amine oxide was not an effective auxiliary for replacement of the amine. A representation of the general reaction is given in Scheme 1.

Table 2 *Yield of 2 by Hydrogen Peroxide Oxidation of 1 Catalysed by Na$_2$MoO$_4$ with Auxiliaries (5 Mole %)*

Auxiliary	% conversion	% yield	Comment
TDA-1	100	92	
Triethylamine	100	71	Unidentified by-product
Tri-n-butylamine	100	92	
Pyridine	<5	~0	
Poly(ethyleneglycol)	<5	<5	slow reaction

Table 3 *Yield of Sulphones Using Hydrogen Peroxide Oxidation Catalysed by Na$_2$MoO$_4$ and Tri-n-butylamine^a*

Thioether	Sulphone % Yield	% Yield Without Catalyst
	98	20
	93	2
	99^b	no reaction
	94	15
	99	89 (sulphoxide)
	95	28 (sulphone) 36 (sulphoxide)

a) Neat at 60°C with Na$_2$MoO$_4$ and tri-n-butylamine (5 mole%). b) Toluene used as solvent.

Organic Phase (neat)

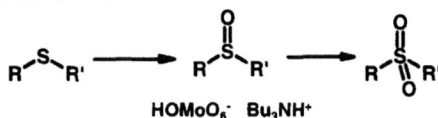

Aqueous Phase (pH3)

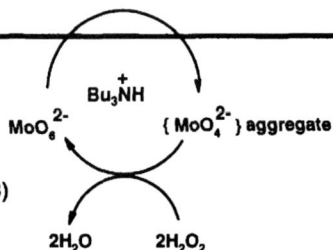

Scheme 1 *Representation of the oxidation of thioethers with hydrogen peroxide catalysed by sodium molybdate and tri-n-butylamine*

3 SUMMARY

The need to bring new Fine Chemical products, with limited lifetimes, to market quickly, and the requirement to meet registered quality, limits the time available for manufacturing technology design and selection. Green technology needs to be capable of rapid evaluation and implementation and be integrated into the total business process.

Trialkylamines can be effectively used as transfer agents in phase transfer peroxomolybdate oxidations of thioethers to sulphones. Use of tri-n-butylamine in this way brings a number of advantages over traditional quaternary ammonium ions; avoidance of tars and easy separation of products from catalyst and auxiliaries.

References

1.   H. B. Freyermuth, H. S. Shultz and S. R. Buc, *J. Org. Chem.*, 1963, **28**, 1140; H. B. Freyermuth, H. S. Shultz and S. R. Buc, *US Patent* 3 005 852.
2.   Z. Stec, J. Zawadiak, A. Skinski and G. Pastuch, Polish J. Chem., 1996, **70**, 1121. D. Bortoloni, F. Di Furia, G. Modena and R Seraglia, *J. Org. Chem.*, 1985, **50**, 2688; N. M. Gresley, W. P. Griffith, and B. C. Parker, *J. Mol. Catalysis (A) Chem.*, 1997, **117**, 185; V. Conte and F. Di Furia, *Catalytic Oxidation with Hydrogen Peroxide as Oxidant,* ed. G. Strukul, Pub. Kluwer Academic, 1992, **7**, 223.
3.   L. J. Csany and K. Jaky, *J. Mol. Catalysis,* 1990, **61**, 75.
4.   G. Soula, *J. Org. Chem.*, 1985, **50**, 3717.
5.   J. P. Ferris, R. D. Gerwe and G. R. Gapski, *J. Org. Chem.*, 1968, **33**, 3493.

NEW CATALYSTS FOR OLD REACTIONS

D. C. Braddock, A. G. M. Barrett, D. Ramprasad and F. J. Waller

Department of Chemistry
Imperial College
South Kensington
London SW7 2AY
e-mail: c.braddock@ic.ac.uk

1 INTRODUCTION

Nitration of aromatic compounds is an immensely important industrial process.[1] The nitroaromatic compounds so produced are themselves widely utilized and act as chemical feedstocks for a great range of useful materials such as dyes, pharmaceuticals, perfumes and plastics. Unfortunately nitrations typically require the use of potent mixtures of concentrated or fuming nitric acid with sulfuric acid leading to excessive acid waste streams and added expense. Alternatively, nitric acid may be used in conjunction with strong Lewis acids such as boron trifluoride.[2] The Lewis acid is used at or above stoichiometric quantities and is destroyed in the aqueous quench liberating large amounts of strongly acidic by-products. With chemists under increasing pressure to perform atom economic processes,[3] creating minimal or no environmentally hazardous by-products, development of novel catalyst systems that facilitate aromatic nitrations in this manner should be of great importance.[4]

Lanthanides have found increasing use as mild and selective reagents in organic synthesis.[5] In particular, lanthanide(III) triflates[6] have been used to good effect as Lewis acids in Diels-Alder,[7] Friedel-Crafts,[8] Mukaiyama aldol[9] and other[10] reactions. For the Mukaiyama reaction the optimum solvent system was found to be aqueous THF; the catalyst was recycled *via* aqueous work up and used repeatedly with little detriment to rate or yield. The compatibility of lanthanide(III) triflate salts with water and other protic solvents and yet their apparent ability to function as strong Lewis acids is somewhat paradoxical. We sought to harness this water tolerant Lewis acidity and have instigated a program in the area of clean technology using lanthanide(III) triflates for atom economic transformations.[11] Herein we report on the use of catalytic quantities of lanthanide(III)[12] and group IV metal[13] triflates [Ln(OTf)$_3$, Ln=La-Lu; M(OTf)$_4$, M= Hf, Zr] and tris(trifluoromethanesulphonyl)methides ("triflides") [M(CTf$_3$)$_3$; M=Yb, Sc] for the nitration of a range of simple and electron deficient aromatic compounds in good to excellent yield using a stoichiometric amount of 69% nitric acid wherein the only by-product is water. Furthermore the catalysts are readily recycled and re-used by a simple evaporative process.

2 RESULTS AND DISCUSSION

2.1 Ytterbium(III) Triflate as a Recyclable Catalyst for the Nitration of Arenes with Nitric Acid

Our investigations began with the commercially available hydrated ytterbium(III) triflate.[14] A range of simple aromatic compounds, both electron rich (quantified by a negative Hammett coefficient, σ_p^+) and electron poor (positive coefficient), were treated with 1 equivalent of 69% nitric acid in refluxing 1,2-dichloroethane in the presence of 10 mol% ytterbium(III) triflate.[15] Initial work utilized anisole (Hammett coefficient σ_p^+ = -0.78), however this led to extensive polymerisation with the formation of intractable organic tars even at room temperature. Alkyl bearing aromatic compounds (Table 1, entries 2, 6, 8) and naphthalene (Table 1, entry 9) were nitrated smoothly and efficiently in refluxing 1,2-dichloroethane and this represents the effective upper limit in reactivity for the arenes (toluene σ_p^+ = -0.31). The lower reactivity limit was next explored. Biphenyl (entry 3, σ_p^+ = -0.18), and bromobenzene (entry 4, σ_p^+ = +0.15) were nitrated successfully (Table 1) but little success was achieved with benzoic acid (σ_p^+=+0.42), acetophenone (σ_p^+ = +0.47), ethyl benzoate (σ_p^+ = +0.48) or benzonitrile (σ_p^+ = +0.70). No dinitrated products were observed in any case and in accord with this system failed to nitrate nitrobenzene (Table 1, entry 5, σ_p^+ = +0.79) and the lower reactivity limit is set at approximate Hammet values of σ_p^+ = +0.3. In the control experiments with no catalyst, only slow reaction occurred and no more than 10% of nitrated products were observed under these conditions.

Table 1 *Nitration of Aromatics with Catalytic Quantities of Yb(OTf)$_3$[a]*

$$R-C_6H_5 \xrightarrow[\text{(CH}_2\text{Cl)}_2, \text{ reflux}]{\text{10 mol\% Yb(OTf)}_3, \text{ 1 equiv. 69\% HNO}_3} R-C_6H_4-NO_2$$

Entry	Arene	% Conversion[b,c]	% Product Distribution[c]		
			ortho	meta	para
1	Benzene	>95 (95)		n/a	
2	Toluene	>95 (95)	52	7	41
3	Biphenyl	89	38	trace	62
4	Bromobenzene	92	44	trace	56
5	Nitrobenzene	0	-	-	-
6	*p*-Xylene	>95		n/a	
7	*p*-dibromobenzene	8		n/a	
8	*m*-Xylene	>95	4-NO$_2$: 85 2-NO$_2$: 15		
9	Naphthalene	>95	1-NO$_2$: 91 2-NO$_2$: 9[d]		

a All reactions carried out on a 3 mmol scale with 10 mol% ytterbium(III) triflate and 1.0 equivalent of 69% nitric acid in refluxing 1,2-dichloroethane (5 ml) for 12 h; *b* Isolated yields in parenthesis; *c* Determined by GC and/or ^{1}H NMR analysis; *d* **Care**: nitronaphthalenes are potent human carcinogens.

Table 2 *Recycled Ytterbium(III) Triflate for the Nitration of m-Xylene[a]*

Run	Conversion (%)[b]	Mass of Catalyst (mg)[c]
1	89	190 (>100)
2	81	152 (82)
3	90	127 (68)
4	88	115 (62)

a All runs performed with 3 mmol m-xylene, 10 mol% ytterbium triflate (run 1) and 1 equivalent of 69% nitric acid in refluxing 1,2-dichloroethane (5 ml) for 5 h; b Determined by GC analysis. The isomeric ratio of 4- and 2-nitroxylene was unchanged throughout (85:15 respectively); c Mass of ytterbium(III) triflate recovered from each run. The figures in parenthesis indicate the percentage recovery which were not optimised.

It is important to note that the only side product from these nitrations is water. With the additional benefit that the catalyst can be recycled (*vide infra*) this simple methodology represents an efficient and environmentally friendly process.

Kobayashi has demonstrated the feasibility of recycling lanthanide(III) triflates for a range of reactions.[16b,16] Consequently, ytterbium(III) triflate could be recovered from the reaction mixture at the completion of any particular nitration run *via* simple partition work-up and isolated by evaporation of the aqueous phase[15] (evaporation of the organic phase gives the nitroaromatic). The resulting free-flowing white solid was found to have an identical IR spectrum to that of the commercially available material and could be re-used as the catalyst for further nitration runs with no loss in rate or yield or change in isomer distributions. The results for four successive nitrations of m-xylene with recycled ytterbium(III) triflate are shown above (Table 2).

2.2 Screening the Lanthanide(III) Triflates

With a view towards rate optimisation a range of lanthanide(III) triflates were examined as potential catalysts. All were found to exhibit catalytic competence but marked differences were apparent (Table 3).

Inspection of the data reveals a clear inverse correlation (with the exception of a few scattered data points) between the ionic radii of the various tripositive lanthanide ions and the extent of nitration whereby the the heavier congeners are the most effective. Thus lanthanum(III) (Z = 57) triflate gave a 64% conversion of naphthalene to mononitronaphthalenes over 1h, whereas the ytterbium(III) (Z =70) triflate catalysed reaction gave a >95% conversion over the same time period.

2.2.1 Postulated Mechanism and Structural Analysis. The isomer distributions from the nitrations of various arenes (Table 1) are consistent with direct electrophilic attack by nitronium ion or, more probably, a nitronium "carrier" of some description.[1] The inverse correlation of ionic radii (which should more properly be expressed as charge-to-size ratio, Z/r) and catalytic competence (Table 3) is indicative of interplay between the lanthanide ion and nitric acid where evidently an increasing electrostatic interaction leads to greater reactivity. On this basis a working mechanism can tentatively be proposed. Firstly, nitric acid binds to the lanthanide metal *via* displacement of water from its inner co-ordination sphere (the reactions are performed with 69% nitric acid and the catalyst resides predominately in the aqueous phase as judged by solubility studies). The triflate counterions are outer sphere and effectively spectator ions (Scheme 1, eqn 1). The resulting strong polarisation due to the metal results in proton liberation affording a lanthanide bound nitrate species (1) (Scheme 1, eqn 2) and the proton goes on to liberate

Table 3 Effect of Various Ln(OTf)₃ for the Nitration of Naphthalene

$Ln(OTf)_3{}^{a,b}$	Ionic Radii/Å^c	Charge-size (Z/r) ratio	%Conversion d
Control	n/a	n/a	38%
La (57)	1.172	2.56	64
Ce (58)	1.15	2.61	-
Pr (59)	1.13	2.65	73
Nd (60)	1.123	2.67	39
Pm (61)	(1.11)	(2.70)	radioactive
Sm (62)	1.098	2.73	67
Eu (63)	1.087	2.76	>95
Gd (64)	1.078	2.78	82
Tb (65)	1.063	2.82	93
Dy (66)	1.052	2.85	64
Ho (67)	1.041	2.88	93
Er (68)	1.033	2.90	95
Tm (69)	1.020	2.94	-
Yb (70)	1.008	2.98	>95
Lu (71)	1.001	3.00	>95

a All reactions performed on a 3 mmol scale in 5ml refluxing 1,2-dichloroethane with 1.05 equivalents of 69% nitric acid and 10 mol% catalyst; *b* the atomic number is shown in parentheses; *c* taken from *lanthanides in organic synthesis*, T. Imamoto, Academic press, 1994, p. 4; *d* Determined by GCMS: 1-nitro:2-nitronaphthalene were obtained in a 9:1 ratio.

NO_2^+ in the classical manner (Scheme 1, eqn 3). Thus, the experimentally observed correlation between increasing charge-to-size ratios (*i.e.* decreasing ionic radii) and extent of conversion is rationalised by noting that the release of the "catalytic proton" is more facile as the metal becomes more polarising.

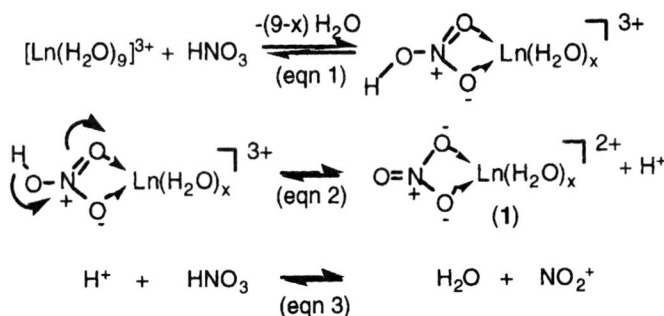

Scheme 1 *Proposed mode of action*

It becomes clear that the lanthanide salt is acting as a "sink" for nitrate ions, displacing the classical equilibrium process (Scheme 2) invoked for the autoionisation of nitric acid to the right hand side; the stronger the binding, the greater the equilibrium concentration of nitronium ion and hence the increased rate of nitration.

$$2HNO_3 \rightleftharpoons NO_3^- + NO_2^+ + H_2O$$

The Lanthanide salt 'captures' nitrate ions increasing the equilibrium concentration of the de facto *nitrating agent: NO_2^+*

Scheme 2 *Effect of lanthanide on classical nitronium ion equilibrium*

Inspection of the proposed nitration mechanism (Scheme 1) reveals that the mononitrate dipositive lanthanide species $[Ln(H_2O)_x(NO_3)](OTf)_2$ (1) is the key intermediate. An independent preparation and characterisation of such a species enables possible indentification of 1 directly *in situ* in the reaction mixture. Additionally, spectroscopic examination of these salts may provide some evidence for our working model. We have developed a novel preparation of these mixed salts by simple metathesis of lanthanide chlorides with the requisite quantities of silver nitrate and silver triflate in water (Scheme 3).[17] The resulting hydrated salts were white or lightly coloured (pink, green or yellow) solids which were found to be stable indefinitely at room temperature in the solid state.

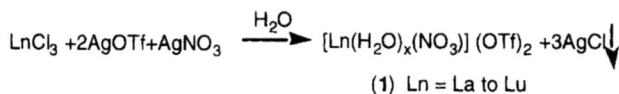

$$LnCl_3 + 2AgOTf + AgNO_3 \xrightarrow{H_2O} [Ln(H_2O)_x(NO_3)](OTf)_2 + 3AgCl\downarrow$$

(1) Ln = La to Lu

Scheme 3 *Preparation of putative intermediate 1*

Characterisation was accomplished by IR spectroscopy. These materials exhibited the six stretches in the IR spectrum indicative of a symmetrically bound (*i.e.* inner sphere) bidentate nitrate species[18] as well as the requisite (outer sphere) triflate stretches. For example, ytterbium(III) based analogue displayed characteristic nitrate absoptions at 1492, 1384, 1326, 814, 769 and 751 cm^{-1}. The stretch at 1492 cm^{-1}, assigned as the A$_1$ symmetrical stretch, is known to be critically dependent on the polarising power of the metal centre where increased polarisation of the nitrate leads to the observation of stretches at increased wavenumbers.[18]

The A$_1$ stretching frequencies were found to increase steadily in magnitude across the lanthanide period for the $[Ln(H_2O)_x(NO_3)](OTf)_2$ salts starting at 1450 cm^{-1} for lanthanum (r^{3+} = 1.172 Å, Z/r = 2.56) and ending at a value of 1497 cm^{-1} for lutetium (r^{3+} = 1.00, Z/r = 3.00) (Table 4). This structural data correlates extremely well with the observed reactivity of the respective lanthanide(III) triflates for nitration and can be taken as strong evidence for the proposed mode of action.

A plot of the A$_1$ IR stretching frequencies versus the charge-to-size ratio of tripositive lanthanide ions reveals two straight lines with an intersection point located around atomic number Z = 60 (Figure 1). This can be interpreted as a change in co-ordination number where the hydration sphere is somewhat more compact for the heavier (and thus smaller) lanthanide ions in line with literature precedent.[19]

Table 4 *Characterisation of [Ln(H$_2$O)$_x$(NO$_3$)](OTf)$_2$ by IR Spectroscopy*

Ln	Ionic Radius (+3) / Å	Charge-to-size Ratio (Z/r)	IR stretch[a] / cm^{-1}
La	1.172	2.56	1450
Ce	1.15	2.61	1461
Pr	1.13	2.65	1469
Nd	1.123	2.67	1477
Pm	(1.11)	(2.70)	(Radioactive)
Sm	1.098	2.73	1478
Eu	1.087	2.76	1482
Gd	1.078	2.78	1484
Tb	1.063	2.82	1480
Dy	1.052	2.85	1490
Ho	1.041	2.88	1489
Er	1.033	2.90	1492
Tm	1.020	2.94	1496
Yb	1.008	2.98	1492
Lu	1.001	3.00	1497

a The bands for these stretches were fairly broad and at times split into two badly resolved signals - the corresponding error due to peak picking is estimated to be *ca.* 5 cm^{-1}.

2.3 Group IV Metal Triflates as Superior Nitration Catalysts

From the above structural analysis it becomes clear that metal triflates with charge-to-size ratios greater than "3" (*i.e.* greater than that of the smallest lanthanide: lutetium) should be more effective nitration catalysts. We considered that the group IV metals hafnium (r^{4+} = 0.78 Å, Z/r = 5.13) and zirconium (r^{4+} = 0.79 Å, Z/r = 5.06) might be suitable for such a purpose. In line with this reasoning we noted that hafnium(IV) triflate has been shown to be an effective catalyst for Friedel-Crafts acylations and alkylations where the corresponding lanthanide salts were less active.[20]

Extrapolation of the plot shown in Figure 1 indicates that a metal cation with a charge-to-size ratio of approximately "5.1" might be expected to show an A$_1$ nitrate stretching frequency in its IR spectrum for salts of the type [M(H$_2$O)$_x$(NO$_3$)](OTf)$_3$ in the vicinity of 1650 cm^{-1} (Figure 2) suggestive of a very tightly bound nitrate and hence a very active nitration catalyst (for M(OTf)$_4$).

Hafnium and zirconium mononitrate tris(triflate), [M(H$_2$O)$_x$(NO$_3$)](OTf)$_3$, were prepared from their tetrachlorides in analogous fashion to the lanthanide salts. Much to our delight these (deliquescent) salts displayed A$_1$ nitrate stretching frequencies in their IR spectra at 1651 and 1650 cm^{-1} respectively. Armed with this pleasing information and with a specific programme aim of nitrating *o*-nitrotoluene (ONT) to dinitrotoluenes (DNTs) with these catalyst types, (catalytic quantities of ytterbium(III) triflate were essentially ineffective for this transformation) hafnium(IV) and zirconium(IV) triflate were prepared as their hydrated salts *via* metathesis of their tetrachlorides with silver triflate in water.

The freshly prepared hafnium salt was employed at a 10 mol% loading for the nitration of ONT with nitric acid in refluxing 1,2-dichloroethane (Scheme 4). After 24h no ONT remained and 2,4- and 2,6- DNT were isolated in 92% yield as a 65:35 mixture

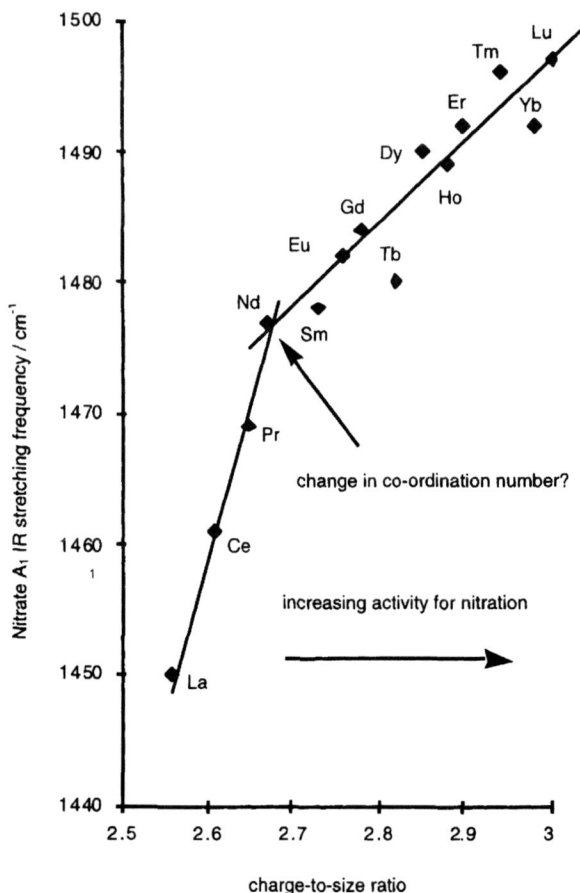

Figure 1 *Plot of A$_1$ IR nitrate stretches for the Ln(H$_2$O)$_x$(NO$_3$)(OTf)$_2$ complexes*

after aqueous work-up. Similarly, zirconium(IV) triflate (10 mol%) was found to have comparable catalytic activity for the nitration of ONT; essentially complete conversion was obtained after 24h, and 2,4- and 2,6-DNT were isolated in 87% yield in a 66:34 ratio.

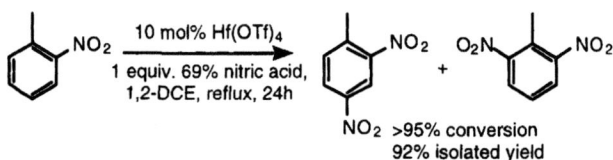

Scheme 4 *Nitration of o-nitrotoluene with Hf(OTf)$_4$*

The catalyst could be recovered by the usual aqueous work-up regime. This recovered material had an identical IR spectrum to that of the freshly prepared catalyst with additional minor signals for (presumably) nitro containing compounds and was utilized

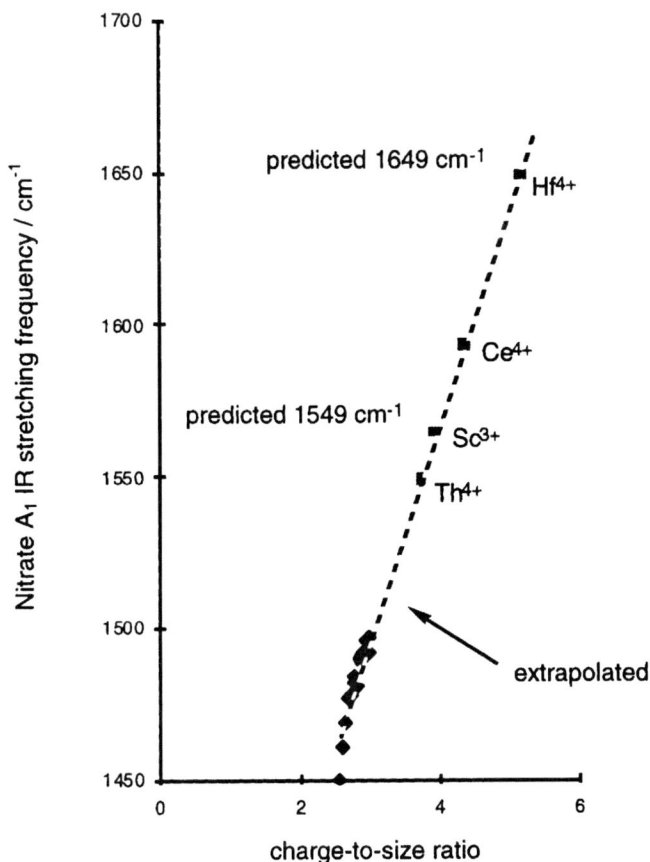

Figure 2 *Extrapolation of A_1 stretching frequencies for metals with increased charge-to-size ratios*

for a futher two nitration runs (Table 5). These results demonstrate that our model based on charge-to-size ratios successfully predicted the activities of various metal triflates for nitrations.

Table 5 *Recycled Zirconium(IV) Triflate for the Nitration of o-Nitrotoluene*

Run[a]	Time / h	% Conversion[b]	Mass Recovery / mg[c]
1	24	> 95 (87)	173 (86)
2	27	> 95 (80)	134 (67)
3	23	88 (76)	121 (61)

a All runs performed with 3 mmol o-nitrotoulene, 10 mol% zirconium(IV) triflate (200 mg) (run 1) and 1 equivalent of 69% nitric acid in refluxing 1,2-dichloroethane (2 ml); b Determined by GC analysis. The isomeric ratio of 2,4- and 2,6-DNT was unchanged throughout (65:35 respectively). The isolated yields are shown in parenthesis; c Mass of zirconium (IV) triflate recovered from each run. The figures in parenthesis indicate the percentage recovery which was not optimised.

2.4 Solvent and Counterion Effects

The two phase nature of the reaction mixture (aqueous nitric acid/lanthanide salt and solvent/substrate) poses a number of questions. Foremost amongst these is the following: in which phase does the actual nitration occur? Comparison of the nitration rates using 1,2-dichloroethane (b.p. 83 °C) versus cyclohexane (b.p. 80 °C) as the solvents (both reactions performed at reflux) allows speculation on this matter. For the nitration of naphthalene with 10 mol% ytterbium(III) triflate a 78% conversion of naphthalene to mononitronaphthalenes occurred over 0.5h in 1,2-dichloroethane whereas for cyclohexane only a 24% conversion was observed. Based on this result it seems reasonable to conclude that the electrophilic substitution process transpires in the organic phase.

We have previously surmised (*vide supra*) that the *de facto* nitrating agent in this system is the nitronium ion. Evidently the nitronium ion diffuses into the organic phase from the aqueous layer where it has been generated. Moreover, it becomes apparent that in order to maintain overall charge neutrality, the nitronium ion must be accompanied by an anionic counterion (*i.e.* triflate). Thus, the triflate conterion fulfills two roles; it is the conjugate base of a sufficiently strong acid such that nitric acid is preferentially protonated, producing the nitronium ion *and* acts as a "shuttle" for the transportation of the nitronium ion into the organic phase where nitration occurs (Scheme 5) and where it is thought that the solubility of the nitronium salt ($NO_2^+OTf^-$) is a key factor.[21]

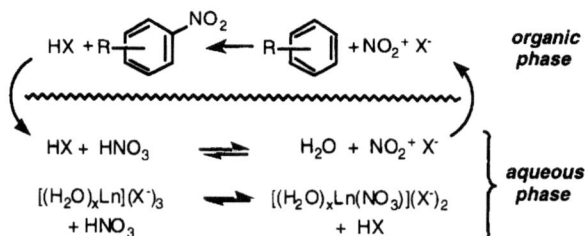

Scheme 5 *The dual rôle of the counterion*

Evidence to support this "counterion modified" model comes from the investigation of hydrated $Ln(NO_3)Cl_2$ prepared in the usual way in water ($LnCl_3$, $AgNO_3$). The lanthanide(III) chlorides themselves show little or no catalytic activity for nitrations. A possible rationale for this is obtained by noting that HCl is a poor activator of nitric acid in nitration chemistry since it is not sufficiently acidic to protonate nitric acid. However, the IR spectra of the $Ln(NO_3)Cl_2$ salts are essentially identical to those of $Ln(NO_3)(OTf)_2$ indicating that the chloride ions are outer sphere in these complexes. Additionally the nitrate bands show the same trend as demonstrated for the triflate series (*e.g.*, for $La(NO_3)Cl_2$ the characteristic nitrate stretch is observed at 1459 cm[-1] and for $Yb(NO_3)Cl_2$ the band appears at 1497 cm[-1]). This indicates that the lanthanide chlorides are capable of activating nitric acid (*via* metal-nitrate interactions) but critically, the counterion (*i.e.* chloride) is incapable of fulfilling its role (in whatever capacity that may be) and hence no nitration occurs.

We considered various other counterions for the lanthanide salts with a view to producing more active nitration catalysts. The use of trifluoroacetate and pentafluorobenzene sulfonate were ineffective presumably on acidity grounds. The use of the classical conjugate bases of strong acids (BF_4^-, PF_6^-, SbF_6^-) was precluded by their known tendency to undergo fluoride abstraction by strong electrophiles and their penchant for hydrolysis.[22]

Yamamoto *et al.* have demonstrated the use of scandium triflamide, $Sc(NTf_2)_3$, for acetylation and acetalisation chemistry where this material is more active than scandium triflate.[23,24] We recognised that Tf_2NH is itself a very strong acid (gas phase acidity measurements have shown it to be a stronger acid than triflic acid[25]) and as such the triflamide anion may well be successful as a counterion for nitrations. Additionally it was hoped that the proposed *in situ* nitrating agent Tf_2N^- NO_2^+ would have enhanced solubility over nitronium triflate thus leading to an additional rate acceleration. Interestingly, Tf_2N^- NO_2^+ is known[26], as is TfO^- NO_2^+,[27] but does not appear to have been utilized for nitrations. Indeed, experiments have shown that catalytic quantities of $Yb(NTf_2)_3$[28] and $Sc(NTf_2)_3$[29] are 2-3 times more effective than $Yb(OTf)_3$ and $Sc(OTf)_3$ for the nitration of toluene respectively. However at this stage it is not clear whether this rate enhancement is due to the triflamides increased inherent acidity or due to a solubility effect or both.

A logical progression along the series TfO^-, Tf_2N^- leads us to Tf_3C^- ("triflide"). Tf_3CH is a known compound.[30] Remarkably, this material has been shown to be an even stronger acid than either HOTf or $HNTf_2$ (in the gas phase)[25] and has been employed as an electrolyte in non-aqueous high voltage batteries,[31] displaying high anodic stability.[32] The Tf_3C^- counterion would seem to be an attractive target on two counts for lanthanide catalysed nitrations: an inherent acidity increase would lead to faster rates and it seems probable that Tf_3C^- NO_2^+ would have increased solubility in organic phases compared to the triflate and triflamide salts.

The literature preparation of the triflide anion is non-trivial involving handling of gases and non-commercial materials. Accordingly, we have recently developed a novel and convenient synthesis of the triflide anion from readily available triflic anhydride and trimethylsilylmethyllithium (Scheme 6).[33]

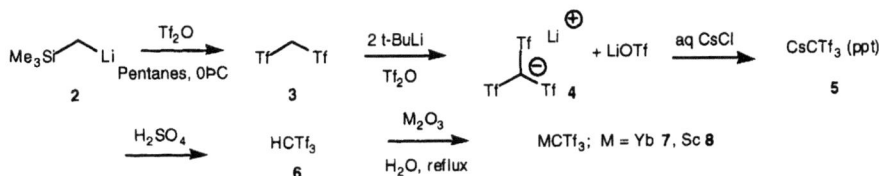

Scheme 6 *Preparation of triflidic acid and ytterbium(III) and scandium(III) triflide*

Slow addition of triflic anhydride to a stirred solution of trimethylsilylmethyllithium (2) in pentanes at 0 °C resulted in smooth reaction as evidenced by the gradual precipitation of lithium triflate with only a slight increase in the internal temperature (max. $\approx$ 10 °C). Basic aqueous quench, extraction with dichloromethane and strong acidification (conc. HCl) of the remaining aqueous phase followed by extraction with dichloromethane led to the isolation of methyleneditriflone (3) as a yellow oil which was essentially pure by ¹H NMR but could be obtained as a low-melting white solid after vacuum sublimation. The acidic nature of 3, which has been estimated to have a pKa of -1[34] allows for easy purification *via* acid-base extraction as above and this procedure allows gram quantities to be prepared in a reproducible 50-55% yield.

Triflylation of 3 with triflic anhydride was carried out by double deprotonation with *tert*-butyllithium followed by the addition of triflic anhydride. The expected reaction product, 4, would, by the precondition of employing triflic anhydride, be contaminated with at least stoichiometric quantities of lithium triflate and a suitable separation regime was required. After quenching with aqueous base, acidification and extraction with dichloromethane to remove all neutral organic side products, lithium triflide and triflate could be extracted effectively from the aqueous phase with diethyl ether. The triflide anion was completely freed from any triflate anion by selective precipitation with cesium chloride in water to give 5. Purification was effected by vacuum sublimation of the cesium salt from concentrated sulfuric acid (generating the free acid (6)) and re-precipitation with cesium chloride followed by recrystallization giving 5 as an analytically pure white solid. The cesium salt (5) is a convenient way to store the triflide anion and does not decompose or color on standing at room temperature. Simple vacuum sublimation from sulfuric acid using a modification of Seppelt's procedure[30] generates the free acid (5) which is best handled as an aqueous solution.

Ytterbium(III) and scandium(III) triflides [$Yb(CTf_3)_3$ (7) and $Sc(CTf_3)_3$ (8)] were obtained by an analogous protocol employed for the triflate series *viz.* the metal oxides were heated to reflux with the free acid in water. Thus, heating at reflux Yb_2O_3 or Sc_2O_3 in an aqueous solution of $HCTf_3$ (6) provided the both the metal triflides and in essentially quantitative yields as white powders after drying for 24h. Recrystallization furnished single crystals of the aqua complexes of both $Yb(CTf_3)_3$ and $Sc(CTf_3)_3$ suitable for X-ray crystallography.

Whilst $Yb(OTf)_3$ was essentially ineffective for the nitration of *o*-nitrotoluene (ONT) to dinitrotoluenes (DNT's) giving 8% conversion under our standard conditions (*i.e.* 10 mol% catalyst, 1 equiv. 69% HNO_3, 1,2-DCE, reflux, 24h), 10 mol% $Yb(CTf_3)_3$ was found to mediate the nitration of ONT to 93% conversion in 24 h. The use of 10 mol% $Sc(OTf)_3$ gave a 50% conversion to DNT's in 24h although the reaction mixture was found to darken considerably. In contrast, $Sc(CTf_3)_3$ was found to give a 91% conversion within the same time period and no darkening was observed. In all cases, the ratio of 2,4-:2,6-DNT'S was found to be 65:35.

In conclusion, we have developed an environmentally friendly nitration procedure with our initial investigations commencing with the commercially available lanthanide(III) triflates. As a result of an in-depth mechanistic study we have delineated the role both of the metal centre and the counterion. We have used this understanding to design superior catalysts by changing the metal centre or the counterion. In all cases the nitration systems use a single equivalent of nitric acid, the only side-product is water and the catalysts may be recycled and reused. We believe this to be a significant step forward in the area of clean technology for aromatic nitration.

Acknowledgements

The Imperial College group thank Air Products and Chemicals Inc. for the support of our research under the auspices of the Joint Strategic Alliance, and the EPSRC.

References and Notes:

1. (a) G. A. Olah, R. Malhotra and S. C. Narang, 'Nitration: Methods and Mechanisms', VCH, New York, 1989; (b) C. K. Ingold, 'Structure and Mechanism in Organic Chemistry', 2nd ed., Cornell University Press, Ithaca, New York, 1969; (c) G. A. Olah and S. J. Kuhn, in 'Friedel-Crafts and Related Reactions', G. A. Olah ed., Wiley-Interscience, New York, Vol. 2, 1964; (d) K. Schofield, 'Aromatic Nitrations', Cambridge University Press, London, 1980.

2. (a) R. J. Thomas, W. F. Anzilotti and G. F. Hennion, *Ind. Eng. Chem.*, 1940, **32**, 408; (b) G. F. Hennion, *U. S. Patent* 2,314,212, 1943.

3. B. M. Trost, *Angew. Chem., Int. Ed. Engl.*, 1995, **34**, 259.

4. For a recent nitration using clean methodology see: K. Smith, A. Musson and G. A. DeBoos, *J. Chem. Soc., Chem. Commun.*, 1996, 469. However this method suffers from the disadvantage of stoichiometric quantities of acidic by-products. For the use of clays and other such solid supported acid sources see Reference 1a.

5. T. Imamaoto, in 'Lanthanides in Organic Synthesis', Academic Press, London, 1994.

6. Reviews: (a) R. W. Marshmann, *Aldrichimica Acta*, 1995, **28**, 77; (b) S. Kobayashi, *Synlett*, 1994, 689; (c) J. B. N. F. Engberts, B. L. Feringa, E. Keller and S. Otto, *Recl. Trav. Chim. Pays Bas*, 1996, **115**, 457.

7. (a) H. Ishitani and S. Kobayashi, *Tetrahedron Lett.*, 1996, **37**, 7357; (b) S. Kobayshi, H. Ishitani, I. Hachiya and M. Araki, *Tetrahedron*, 1994, **50**, 11623; (c) I. E. Marko and G. R. Evans, *Tetrahedron Lett.*, 1994, **35**, 2771; (d) T. Saito, M. Kawamura and J.-I. Nishimura, *Tetrahedron Lett.*, 1997, **38**, 3231.

8. (a) A. Kawada, S. Mitamura and S. Kobayashi, *J. Chem. Soc., Chem. Commun.*, 1996, 183; (b) A. Kawada, S. Mitamura and S. Kobayashi, *J. Chem. Soc., Chem. Commun.*, 1993, 1157.

9. (a) S. Kobayashi and S. Nagayama, *J. Org. Chem.*, 1997, **62**, 232; (b) S. Kobayashi and I. Hachiya, *J. Org. Chem.*, 1994, **59**, 3590 and references cited therein.

10. (a) S. Kobayashi, H. Ishitani and M. Ueno, *Synlett*, 1997, 115; (b) S. Kobayshi, H. Ishitani, S. Komiyama, D. C. Oniciu and A. R. Katritzky, *Tetrahedron Lett.*, 1996, **37**, 3731; (c) J. H. Forsberg, V. T. Spaziano, T. M. Balasubramanian, G. K. Liu, S. A. Kinsley, C. A. Duckworth, J. J. Poteruca, P. S. Brown and J. L. Miller, *J. Org. Chem.*, 1987, **52**, 1017; (d) M. Meguro and Y. Yamamoto, *Heterocycles*, 1996, **43**, 2473; (e) S.-C. H. Diana, K.-Y. Sim and T.-P. Loh, *Synlett*, 1996, 263; (f) M. Meguro, N. Asao and Y. Yamamoto, *J. Chem. Soc., Perkin Trans. 1*, 1994, 2579; (g) R. Annunziata, M. Cinquini, F. Cozzi, V. Molteni and O. Schupp, *J. Org. Chem.*, 1996, **61**, 8293; (h) P. E. Harrington and M. A. Kerr, *Synlett*, 1996, 1047; (i) Y. Makioka, T. Shindo, Y. Taniguchi, K. Takaki and Y. Fujiwara, *Synthesis*, 1995, 801; (j) S.-I. Fukuzawa, T. Tsuchimoto and T. Kanai, *Bull. Chem. Soc. Jpn.*, 1994, **67**, 2227; (k) S. Hosono, W.-S. Kim, H. Sasai and M. Shibasaki, *J. Org. Chem.*, 1995, **60**, 4; (l) M. Chini, P. Crotti, L. Favero, F. Macchia and M. Pineschi, *Tetrahedron Lett.*, 1994, **35**, 433; (m) H. C. Aspinall, A. F. Browning, N. Greeves and P. Ravenscroft, *Tetrahedron Lett.*, 1994, **35**, 4639. (n) S. Matsubara, M. Yoshioka and K. Utimoto, *Chem. Lett.*, 1994, 827; (o) P. G. Cozzi, B. Di Simone and A. Umani-Ronchi, *Tetrahedron Lett.*, 1996, **37**, 1691; (p) G. Jenner, *Tetrahedron Lett.*, 1996, **37**, 3691; (q) T. Hanamoto, Y. Sugimoto, Y. Yokoyama and J. Inanaga, *J. Org. Chem.*, 1996, **61**, 4491; (r) E. Keller and B. L. Feringa, *Synlett*, 1997, 842.

11. For the atom economic acylation of alcohols using acetic acid as the acetyl source where the only side product is water and the catalyst is readily recyclable see: A. G. M. Barrett and , D. C. Braddock, *Chem. Commun.*, 1997, 351.

12. F. J. Waller, A. G. M. Barrett, D. C. Braddock and D. Ramprasad, *Chem. Commun.*, 1997, 613. For an interesting report utilizing $La(NO_3)_3/HCl/NaNO_3$ for the nitration of phenols (but not applicable to any less electron rich aromatics) see: M. Ouertani, P. Girard and H. B. Kagan, *Tetrahedron Lett.*, 1982, **23**, 4315.

13. F. J. Waller, A. G. M. Barrett, D. C. Braddock and D. Ramprasad, *Tetrahedron Lett.*, 1998, **39**, 1641.

14. All the lanthanide(III) triflates are commercially available from the Aldrich Chemical Co. bar Promethium (radioactive) and Cerium (available in its +4 oxidation state).

15. The following procedure is representative: Nitric acid (69%; 192 ml, 3.0 mmol) was added to a stirred suspension of ytterbium(III) triflate (186 mg, 0.30 mmol) in 1,2-dichloroethane (5 ml). The suspension dissolved to give a two phase system in which the aqueous phase was the more dense. Toluene (240 ml, 3.0 mmol) was added and the stirred mixture was heated at reflux for 12 h. During the reaction a white solid precipitated and the organic phase became yellow, and after 12 h no phase boundary was apparent. The solution was allowed to cool and diluted with water. The yellow organic phase was dried ($MgSO_4$) and evaporated to give nitrotoluene (390 mg, 95%). The colourless aqueous phase was evaporated to give ytterbium(III) triflate as a white free-flowing solid (183 mg, 98%).

16. (a) S. Kobayshi, I. Hachiya and Y. Yamanoi, *Bull. Chem. Soc. Jpn.*, 1994, **67**, 2342; (b) S. Kobayshi, I. Hachiya and T. Takahori, *Synthesis*, 1993, 371; (c) S. Kobayshi, I. Hachiya, T. Takahori, M. Araki and H. Ishitani, *Tetrahedron Lett.*, 1992, **33**, 6815.

17. The following procedure is representative: ytterbium(III) chloride (775 mg, 2 mmol) as a solution in water was added to a solution of silver nitrate (340 mg, 2 mmol) and silver triflate (1.03 g, 4 mmol) in water. A white precipitate (AgCl) was formed immediately which was filtered at the pump giving a colourless solution. The solution was evaporated under reduced pressure to give a white solid (1.36 g, 99%).

18. C. C. Addison, N. Logan and S. C. Wallwork, *Quarterly Rev.*, 1971, **25**, 289.

19. (a) A. Chatterjee, E. N. Maslen and K. J. Watson, *Acta Cryst.*, 1988, **B44**, 381; (b) T. Lu, L. Ji, M. Tan, Y. Liu and K. Yu, *Polyhedron*, 1997, **16**, 1149; (c) L. I. Semenova, B. W. Skelton and A. H. White, *Aust. J. Chem.*, 1996, **49**, 997; (d) D. L. Faithfull, J. M. Harrowfield, M. I. Ogden, B. W. Skelton, K. Third and A. H. White, *Aust. J. Chem.*, 1992, **45**, 583; (e) J. M. Harrowfield, W. M. Lu, B. W. Skelton and A. H. White, *Aust. J. Chem.*, 1994, **47**, 321.

20. I. Hachiya, M. Moriwaki and S. Kobayshi, *Bull. Chem. Soc. Jpn.*, 1995, **68**, 2053.

21. Nitronium triflate is approximately 150-fold more reactive than the "more-ordered" nitronium tetrafluoroborate and hexafluorophosphate salts in chlorinated solvents and this is attributed to increased solubility conferred to the nitronium ion by the triflate counterion. (Ref. 27a)

22. S. H. Strauss, *Chem. Rev.*, 1993, **93**, 927.

23. (a) K. Ishihara, M. Kubota and H. Yamamoto, *Synlett*, 1996, 265; (b) K. Ishihara, Y. Karumi, M. Kubota and H. Yamamoto, *Synlett*, 1996, 839.

24. For the use of bis(trifluoromethanesulfonyl)imide as a alternative counterion in conjunction with lanthanides, see: (a) H. Kobayashi, J. Nie and T. Sonada, *Chem.*

Lett., 1995, 307; (b) K. Mikami, O. Kotera, Y. Motoyama, H. Sakaguchi and M. Maruta, *Synlett*, 1996, 171; For the use of perfluorooctanesulfonate see: (c) T. Hanamoto, Y. Sugimoto, Y. Z. Jin and J. Inanaga, *Bull. Chem. Soc. Jpn.*, 1997, **70**, 1421. For the use of a sulphonated resin as a 'counterion' see: (d) L. Yu, D. Chen, J. Li and P. G. Wang, *J. Org. Chem.*, 1997, **62**, 3575.

25. I. A. Koppel, R. W. Taft, F. Anvia, S-Z. Zhu, L.-Q. Hu, K.-S. Sung, D. D. DesMarteau, L. M. Yagupolskii, Y. L. Yagupolskii, N. V. Ignat'ev, N. V. Kondratenko, A. Y. Volkonskii, V. M. Vlasov, R. Notario and P.-C. Maria, *J. Am. Chem. Soc.*, 1994, **116**, 3047.

26. J. Foropoulos Jr. and D. D. DesMarteau, *Inorg. Chem.*, 1984, **23**, 3720.

27. (a) C. L. Coon, W. G. Blucher and M. E. Hill, *J. Org. Chem.*, 1973, **38**, 4243; (b) L. M. Yagulpol'skii, I. I. Maletina and V. V. Orda, *Zh. Org. Khim.*, 1974, **10**, 2226; (c) F. Effenberger and J. Geke, *Synthesis*, 1975, 40.

28. The ytterbium salt was prepared from commercially availlable lithium triflamide (3M chemical company). A representative procedure is given below (Ref. 29)

29. $Sc(NTf_2)3.8H_2O$ was prepared as follows: A column (2cm diameter and 21 cm in length) was loaded with an aqueous slurry of Amberlyst A-26 in the chloride form. Lithium triflamide (48.8g, 0.166 mol) dissolved in 1100 ml of water was eluted through the column. A halide test was negative in the last collected fractions. The column was washed with 50 0ml water followed by 200 ml methanol. Scandium triflate (0.9g, 1.82 mmol) was dissolved in 100 ml methanol and eluted through the column. The column was washed with 100 ml methanol and all washings were collected and evaporated in air to yield an oil. Under vacuum, the oil became a light tan solid (1.5 g). A ^{13}C NMR in D_2O showed a characteristic quartet. A ^{19}F NMR spectrum showed a singlet at 79.29 ppm. The solid had 15% water by Karl-Fischer analysis. The best fit from all the data is $Sc[N(SO_2CF_3)_2.8H_2O$ with a theoretical Sc of 4.37% (found 4.97%).

30. L. Turowsky and K. Seppelt, *Inorg. Chem.*, 1988, **27**, 2135.

31. (a) L. A. Dominey, *U. S. Patent* 5,273,840, 1993; (b) L. A. Dominey, V. R. Koch and T. J. Blakley, *Electrochim. Acta*, 1992, **37**, 1551; D. Benrabah, D. Baril, J. Y. Sanchez, M. Armand and G. G. Gard, *J. Chem. Soc., Faraday Trans.*, 1993, **89**, 355; (c) V. R. Koch, C. Nanjundiah, G. Appetecchi, G. Battista and B. Scrosati, *J. Electrochem. Soc.*, 1995, **142**, L116; (d) D. Aurbach, O. Chusid, I. Weissman and P. Dan, *Electrochem. Acta*, 1996, **41**, 747.

32. (a) V. R. Koch, L. A. Dominey, C. Nanjundiah and M. J. Ondrechen, *J. Electrochem. Soc.*, 1996, **143**, 798; (b) C. W. Walker Jr., J. D. Cox and M. Salomon, *J. Electrochem. Soc.*, 1996, **143**, L80.

33. F. J. Waller, A. G. M. Barrett, D. C. Braddock and D. Ramprasad, R. M. McKinnell, J. P. White, D. J. Williams and R. Ducray, *J. Org. Chem.*, 1999, **64**, 2910.

34. R. J. Koshar and R. A. Mitsch, *J. Org. Chem.*, 1973, **38**, 3358.

CATALYSIS FOR FINE CHEMICALS : AN INDUSTRIAL PERSPECTIVE

Pascal Métivier

Rhodia
Centre de Recherches de Lyon
85 avenue des Frères Perret
69192, Saint Fons Cedex
France

1 INTRODUCTION

1.1 Business Considerations

Research and development in the field of fine chemicals differs considerably from bulk chemicals. Requirements for industrial success of a new process are different from bulk chemicals where the most important feature is cost performances. Although this feature is of course important for development of fine chemicals, two other main parameters should be considered. First of all, time to market: One must be ready to manufacture the product at the right time and for a limited period of time. The lifetime of most fine chemicals is much shorter than for bulk chemicals where 20 to 50 years is standard. Second, possible R&D expenses are much lower than for bulk chemicals (Figure 1).

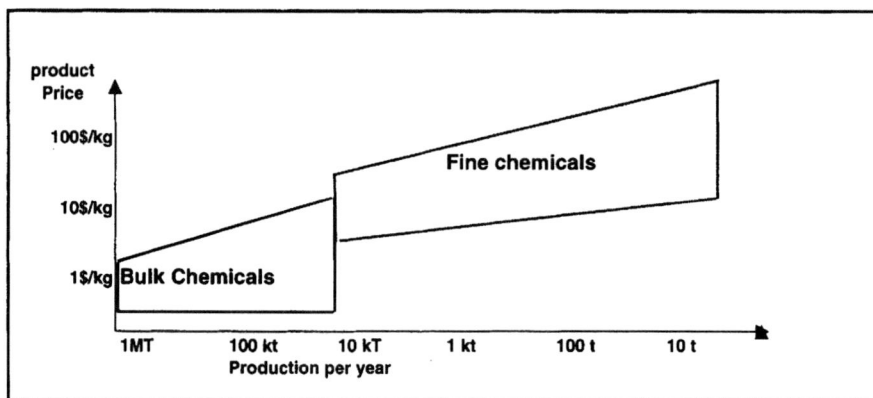

Figure 1 *Compared prices and production characteristics of fine chemicals versus bulk chemicals*

Very few fine chemicals products can be treated as bulk chemicals, nevertheless a few products such as vanillin (aroma), menthol (perfume), ibuprofen (pharmaceutical)...

typically utilise the bulk chemicals R&D methodology, because of their high overall turnover and of their probable long lifetime (Scheme 1).

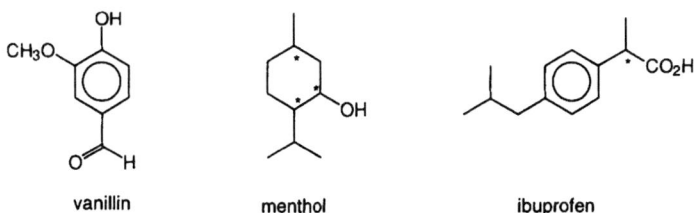

vanillin   menthol   ibuprofen

Scheme 1 *Examples of fine chemicals that can be « identified » as bulk chemicals*

Most of the new developments in fine chemicals will come from small turnover products with short lifetimes and so short R&D development times and low investment possibilities (Figure 2).

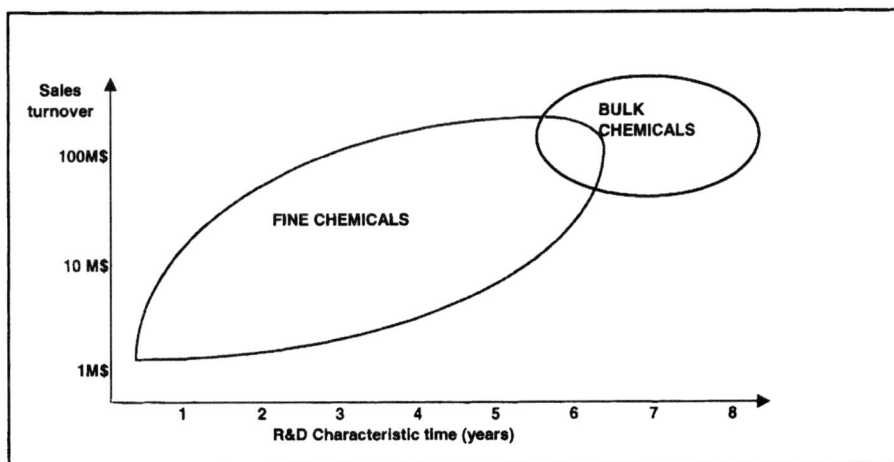

Figure 2 *Compared lifetimes and sales turnovers of fine chemicals versus bulk chemicals*

1.2 Scientific Considerations

Partly as a result of business constraints on research, many fine chemical processes are a long way from optimum, especially when compared to bulk chemical processes, in terms of yield, productivity and effluents. Important improvements in competitiveness could be achieved by development of new catalysts. Due to complexity, development of new catalysts generally requires time, and the industry generally prefers the use of old classical methods that are well known and require a short development time. The three technical requirements for a new catalytic process to be industrialised are the following: 1) The chemistry must be already well advanced at the start of the project so as to fit with the timetable, 2) The chemistry must be easily adaptable to the existing industrial plant (with little modifications), 3) The chemistry must provide a breakthrough in terms of economics and effluents compared to existing processes.

The industrial approach chosen by Rhodia is to develop new technologies applicable to a wide range of products so that chemical feasibility is already demonstrated when the market need for the product arises. Careful choice of reactions to study and correct evaluation of breakthrough is key.

Examples of this type of research will be exemplified with different types of reactions. For each reaction, its interest will be discussed as well as the technology that has been developed. The different reactions that will be discussed are the following :

- Friedel-Crafts reactions : what possible technology might replace the old stoicheiometric Lewis acid route ? The use of acido/basic (particularly zeolites) materials to catalyse this reaction will be discussed as well as the industrial aspects such as recycling, regeneration, and process simplification.
- Carbonylation : How to avoid the use of CO to perform this extremely versatile reaction ? The use of carbonylation has been very limited up to now because of the use of « non friendly » CO reagent. Replacement of this reagent will be discussed as well as the classical issues of recycling regarding homogeneous chemistry.
- Reduction of acids to aldehydes : one step from the acid, how to avoid the Rosenmund reaction or the reduction to alcohol followed by oxidation ? Aldehydes are very important reagent for fine organics, the access to those compounds is always multistep. Development of a catalyst for the reduction of an acid function to the aldehyde without reduction of a double bond in the framework of the molecule will be presented and discussed.
- Oxidation of alcohol's and ketones : Replacement of undesirable reagents such as Jones, permanganate or hypochlorite ? Oxidation of alcohols is one of the most used oxidation procedures, nevertheless, in most cases stoicheiometric, non friendly reagents are required to perform these reactions. Heterogeneous solutions developed in Rhodia using air or oxygen and classical catalysts will be presented and discussed.

2 THE FRIEDEL-CRAFTS ACYLATION OF AROMATICS

The Friedel-Crafts reaction as well as the related Fries rearrangement of aromatics are the most important methods in organic chemistry for synthesizing aromatic ketones,[1] which are important intermediates for the production of fine chemicals (Scheme 2).[2]

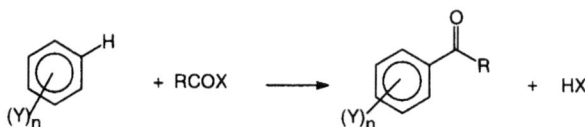

Scheme 2 *Friedel-Crafts reaction*

The conventional method for preparation of these aromatic ketones involves reaction of the aromatic hydrocarbon with a carboxylic acid derivatives using a Lewis acid ($AlCl_3$, $FeCl_3$, BF_3, $ZnCl_2$, $TiCl_4$) or Bronsted acids (polyphosphoric acid, HF). The major drawback of the Friedel-Crafts reaction is the need to use a stoicheiometrical quantity of Lewis acid relative to the formed ketone. This quantity is required due to the fact that the ketone (product of the reaction) forms a stoicheiometrical stable complex with the Lewis acid. The decomposition of this complex is generally carried out with water, leading to total destruction and loss of the Lewis acid.

The Friedel-Crafts reaction was discovered at the end of the nineteenth century, and despite intensive studies a general solution to this initial drawback has not been found.[3] Studies have been carried out to find mild Lewis acids forming less stable complexes with the ketone and still able to catalyze the reaction. Some successes have been obtained using rare earth triflate[4] or bismuth III salts.[5] These methods, if they proved to be catalytic, still require significant quantities of catalysts (a few percent) and the recycling of the catalyst is not simple.

Heterogeneous catalysis appears to be an interesting alternative technique to the homogeneous reaction. The physical and chemical competitive adsorption parameters can in principle be tuned so as to favor displacement of the product by one of the reagents; reactivity can be promoted using « shape » factors to compensate for the use of mildly acidic catalysts. Important work has been carried out in the past 10-15 years to find new heterogeneous catalysts for this reaction.

Systematic work was then undertaken in the Rhodia company[6] to try to rationalize the observed effects. Results obtained using the same reaction conditions and changing only the catalyst are depicted in Table 1.

All tested catalysts are active in this reaction and they all show a very high para selectivity.

The Rhodia company is operating an industrial process for acylation of anisole to para-acetoanisole using zeolite.[7] This process using a fixed bed technology is a breakthrough in this field. It allows a considerable simplification of the process as well as an increase in para selectivity, and so a reduction of its operating cost and a dramatic reduction of the effluents. Figures 3 and 4 illustrate the simplification brought by the zeolite process just comparing the two block diagrams.

This zeolite process is a good example of how a new process can at the same time be cost efficient and environmentally friendly. This technology has now been adapted to the industrial synthesis of acetoveratrole using the same type of process.

Table 1 *Acetylation of Anisole with Acetic Anhydride Using Various Heterogeneous Catalysts (8 h, 90°C)*

Entry	Catalyst	Yield (%)[a]	Selectivity (para)
1	HZSM5	12	> 98
2	H Mordenite	29	> 98
3	$H\beta$	70	> 98
4	HY	69	> 98
5	exchanged clay	14	> 98
6	Al clay	16	> 98
7	$H_2PW_6Mo_6O_{40}$	21	> 98

a : yield on acetic anhydride (initial ratio anisole / anhydride) = 5

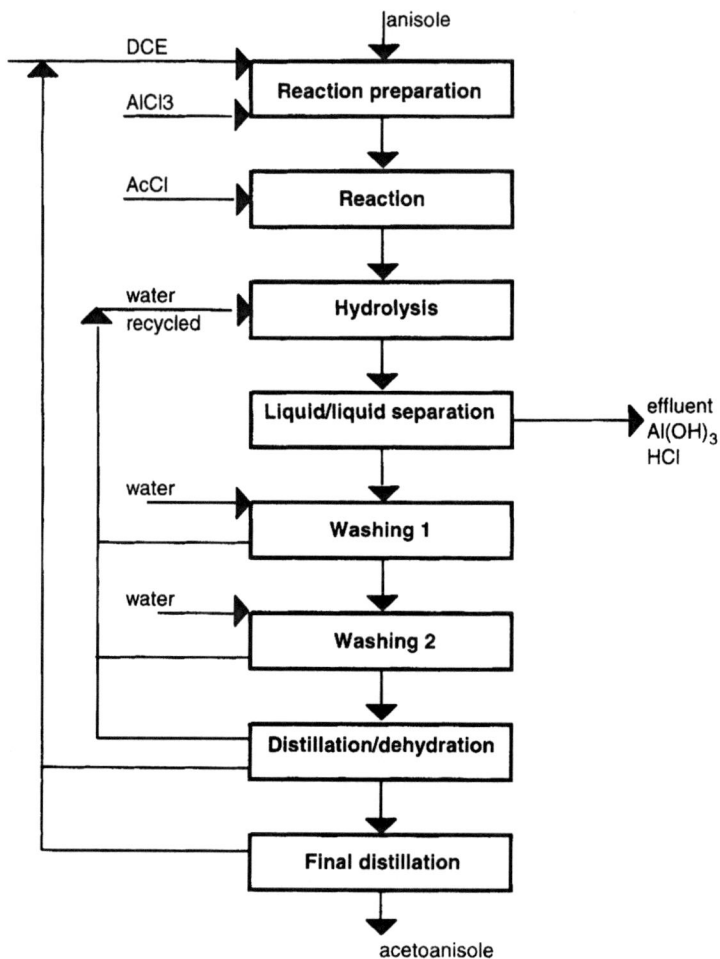

Figure 3 *Simplified flow chart for the conventional aluminum dichloride process using DCE (1,2 dichloroethane) as solvent*

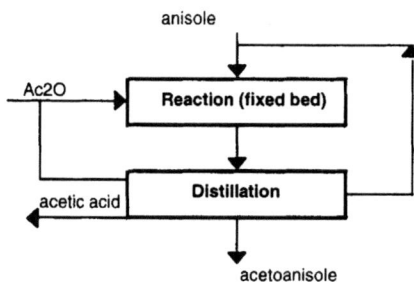

Figure 4 *Simplified flow chart for the new zeolite fixed bed process using no solvent*

3 CARBONYLATION WITHOUT THE USE OF CO

Carbon monoxide chemistry has been extensively studied, leading to a wide range of methods used in small scale organic syntheses up to industrial processes.[8] Despite the versatility of carbonylation reactions, carbon monoxide suffers from major drawbacks that restricts its utilisation. From an industrial point of view, the cumbersome handling of this toxic gas necessitates very expensive facilities which prevent its use for the majority of fine chemical production processes. An alternative process equivalent to a carbonylation reaction which avoids carbon monoxide introduction into the reactor and that can be used in standard polyvalent type units would be of great interest. Of course, catalyst cost, stability and productivity should also fulfil economic requirements.

Previous carbonylation studies, indicated that iridium based catalysts are active in carbonylation under low CO partial pressure. Furthermore these catalysts are active for isomerisation of methyl formate with no initial CO pressure, but precipitate during the reaction.[9] After more detailed investigations of formate isomerisation processes we found that the hydroxycarbonylation of any carbonylable function with formic acid using iridium based catalysts was possible without the use of CO gas. Alkenes and alcohols can be transformed to the corresponding hydroxycarbonylated products with good yields. *n*-Hexene is carbonylated in 92% yield with a standard selectivity to linear acids neighbouring 70%. We compared this reaction with the hydroxycarbonylation of *n*-hexene with CO gas using the same catalyst and found similar results (Table 2).

The scope of this reaction appears to be very great and examples are depicted in Scheme 3.

Two possible mechanism pathways may be involved for this reaction, either a common pathway for all reactants passing through the alkene formation followed by a « standard » hydroxycarbonylation of the substrate with CO formed in situ by decomposition from formic acid, or through oxidative insertion of iridium to an iodoakyl intermediate, corresponding to the carbonylation mechanism of an alcohol. These two possibilities are depicted in Scheme 4.

4 REDUCTION OF ACIDS TO ALDEHYDES

Access to the aldehyde function has always been very important in organic chemistry. The two main transformations to introduce this function are oxidation of a primary alcohol to the aldehyde or reduction of an acid derivatives to the aldehyde. This second reaction has not found any general simple solution. If some progress has been made, the main methods used today are either the Rosenmund[10] reaction on the acid chloride, reduction of the ester using a « sophisticated » hydride, or reduction of the ester to the alcohol followed by re-oxidation of the alcohol function to the aldehyde.

Table 2 *Hydroxycarbonylation of n-Hexene*

Entry	Substrate	CO source	Pressure [bar]	Yield [%]
1	*n*-hexene	HCOOH	Autogeneous	92 (68)
2	*n*-hexene	gas	5	86 (63)

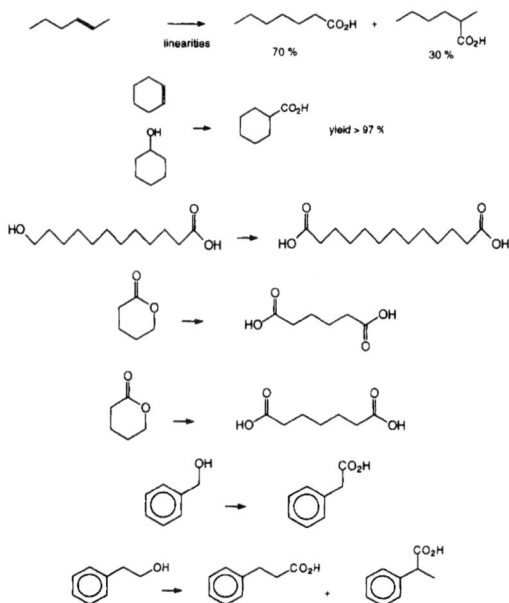

Scheme 3 *Examples of reactions performed using formic acid and an iridium based catalysis*

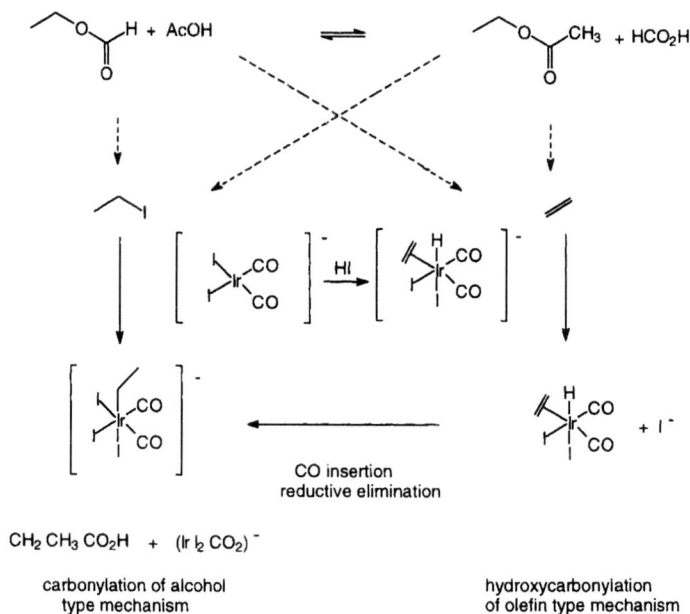

Scheme 4 *Possible mechanisms using the formic acid/iridum catalysed procedure*

Table 3 *Examples of Results Obtained for Reduction of Acids to Aldehydes Using Ru/Sn Type Catalyst*

Reagent	Product	Conversion	Selectivity
 Senecioic acid	 *prenal*	40%	80%
CF_3CO_2H *trifluoro acetic acid*	CF_3CHO *fluoral*	80%	90%
 salicylic acid	 *salicylic aldehyde*	65%	65%
 nonanoic acid	 *nonanal*	85%	90%
 undecylenic acid	 *undecenal*	85%	68%

We decided to look for a « more simple » catalytic system able to reduce an acid function into an aldehyde function. Moreover, we had identified selective reduction of an acid using hydrogen, without reduction of a double bond situated in the framework of the molecule, as a breakthrough target. Screening of a large variety of catalysts in different conditions, lead to the selection of one active and selective catalyst and this using gas phase technology. A Ru/Sn supported type catalyst can be very active and selective toward this reaction.[11] Results obtained with this type of catalyst are depicted in Table 3.

Careful investigation of the catalyst indicates that the active phase is a Ru_3Sn_7 alloy phase. Ru_3Sn_7 is a cubic structure of the Im3m spatial group, which contains tin antiprisms as characteristic structure.

5 OXIDATION OF ALCOHOLS

Oxidation of alcohols to aldehydes or ketones has always been considered as a difficult reaction by organic chemists, they generally prefer to avoid the use of oxidation reaction during a synthesis. Most of the common reactions involve stoicheiometric, environmentally unfriendly reagents with names such as Jones, Collins, and others.[12] In Rhodia we perform on an industrial basis the oxidation of an alcohol to the corresponding aldehyde using oxygen as the reagent and a precious metal promoted with Bi III salts as the catalyst[13] (Scheme 5). This industrial reaction is carried out in batch conditions and could be adapted to a wide variety of products. We decided to investigate the use of such reaction for difficult cases that could arise.

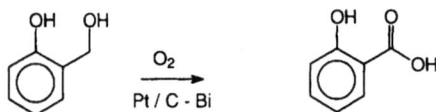

Scheme 5 *Oxidation of a benzylic alcohol*

First careful studies on benzylic alcohols indicates that it is possible to oxidise stepwise with high selectivity to the aldehyde and then to the ketone. For example oxidation of bis-hydroxymethylguaiacol (precursor for trimethoxybenzaldehyde) can be selectively stopped at the bis-aldehyde stage or pursued to the carboxyvanillin as depicted in Scheme 6.[14]

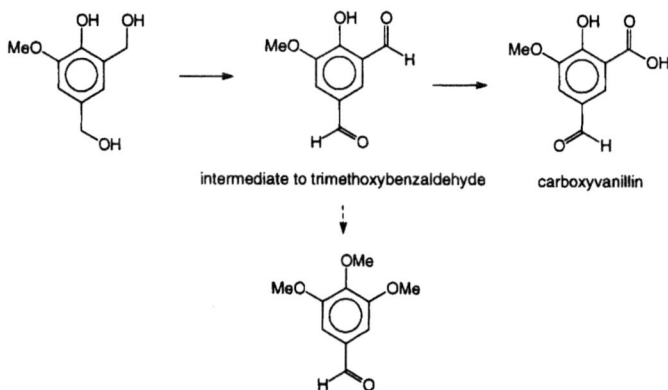

intermediate to trimethoxybenzaldehyde   carboxyvanillin

Scheme 6 *Example of selective oxidations that can be obtained*

This ortho to para selectivity has been extended to intramolecular reaction, and using this technique ortho-vanillin can be selectively oxidised to ortho-vanillic acid in the presence of para vanillin[15] (Scheme 7).

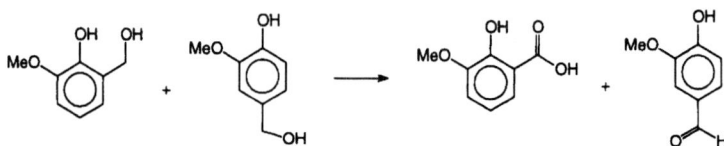

Scheme 7 *Example of intermolecular selectivity using Pd or Pt on charcoal promoted with Bi catalysts*

6 CONCLUSIONS AND PERSPECTIVES

Research to develop new specific catalysts for fine chemicals must be applicable to a range of products and cannot be limited to only one compound. Furthermore, since development time is limited, the chemical feasibility of the reaction must be demonstrated in advance. Since synthesis of fine chemicals has not been fully studied up to now, there is a unique opportunity to review classical organic chemistry and to find and develop new selective catalysts for key reactions. Since the reactions must be general careful choice of reactions to investigate is key to success. We have demonstrated that for important reactions such as Friedel-Crafts, Carbonylation, Reductions and oxidations, it is possible to develop new catalysts for the selective synthesis of fine chemicals.

References

1. G. A. Olah, ' Friedel-Crafts and Related Reactions', Wiley Interscience, New York, 1963-1964, **Vol I-IV**; G. A. Olah ' Friedel-Crafts Chemistry', Wiley Interscience, New York, 1973.
2. H. Szmant, 'Organic Building Blocks of the Chemical Industry', Wiley, New York, 1989.
3. C. Friedel and J. M. Crafts, *Bull. Chem. Soc. Chim. Fr.*, 1877, **27**, 482; C. Friedel and J. M. Crafts, *Bull. Chem. Soc. Chim. Fr.*, 1877, **27**, 530.
4. A. Kawada, S. Mitamura and S. Kobayashi, *J. Chem. Soc., Chem. Commun.*, 1993, 1157; L. Hachiya, M. Mariwaki and S. Kobayashi, *Tetrahedron Lett.*, 1995, **36**, 409.
5. J. R. Desmurs, M. Labrouillere, J. Dubac, A. Laporterie, H. Gaspard and F. Metz, in 'Industrial Chemistry Library', Elsevier, 1996, **8**, 15-25.
6. M. Spagnol, L. Gilbert and D. Alby, in 'Industrial Chemistry Library', Elsevier, 1996, **8**, 29-38.
7. M. Spagnol, L. Gilbert, E. Benazzi and C. Marcilly, Patent to Rhodia, 1996, WO 96/35655
8. H. M. Colquhoun, D. J. Thompson, and M. V. Twigg, in 'Carbonylation : Direct Synthesis of Carbonyl Compounds', Plenum Press, New York, 1991.
9. C. Patois, R. Perron and D. Thiebaut, Patent to Acetex Chimie, priority 27/03/1996, WO 97/35828.
10. H. van Bekkum and M. Peters, *Recl. Trav. Chim. Pays-Bas*, 1971, **90**, 1923; Burgstahler, Weigel, and Shaefer, *Synthesis*, 1976, 767.
11. R. Jacquot and R. M. Ferrero, Patent to Rhodia, priority 24/11/1991, EP 0539274; R. Jacquot, Patent to Rhodia, priority 08/11/1995, EP 0874687.
12. J. March, in 'Advanced Organic Chemistry', 4th Edition, Wiley-Interscience, New York, 1992, 1167.
13. J. Le Ludec, Patent to Rhodia, priority 07/10/1976, DE 2612844.
14. P. Métivier, Patent to Rhodia, priority 24/05/1995, WO 96/32454.
15. P. Denis, C. Maliverney and P. Métivier, Patent to Rhodia, priority 14/10/1996, WO 98/16493.

A CASE STUDY ON RECOVERY AND REUSE OF COMPLEX SOLVENT MIXTURES FROM CHEMICAL PRODUCTION

Ted Lee

Novartis Ringaskiddy Limited,
Ringaskiddy,
Co. Cork, Republic of Ireland

1 INTRODUCTION

Complex multiphase process waste streams are generated in the production of active substances in the chemical industry. A number of standard types of chemical reactions currently carried out in the pharmaceutical industry would generate complex waste solvent mixtures containing all or some of the following components, tetrahydrofuran, hexane, alcohols, esters and water. The aim of this project was to demonstrate the feasibility to develop processes and to install a recovery plant which separates the multiphase waste streams from the production of an active drug substance into their original components and purify them for reuse. The project was successful in achieving its aims. The processes developed support sustainable development in industrial activities and demonstrate an integrated approach to the environment and industry.

2 PROJECT BACKGROUND

A general process flow diagram for manufacturing bulk drug substances at Novartis Ringaskiddy is outlined in Figure 1.

The waste streams from production can be large, complex and multiphase. A study was undertaken to investigate the possibilities to reduce the volume of solvent required and to recover and reuse the waste solvents from one process. This approach is in accordance with the hierarchy of preferred options for handling waste (see Figure 2) and in line with the European Union's Fifth Action Programme.

Processes were developed at laboratory scale to separate and purify the large multiphase solvent waste streams into their original components for reuse.

The next step was to demonstrate, on a large scale, the feasibility of the processes developed at laboratory scale to separate and purify the waste solvent streams into their original components for reuse. An offer of co-financing was received from the European Union's LIFE -- environment. LIFE (Financial Instrument for the Environment) is a financial instrument used by the European Commission to support the development and implementation of the Community environmental policy as described in their Fifth Community Action Programme.

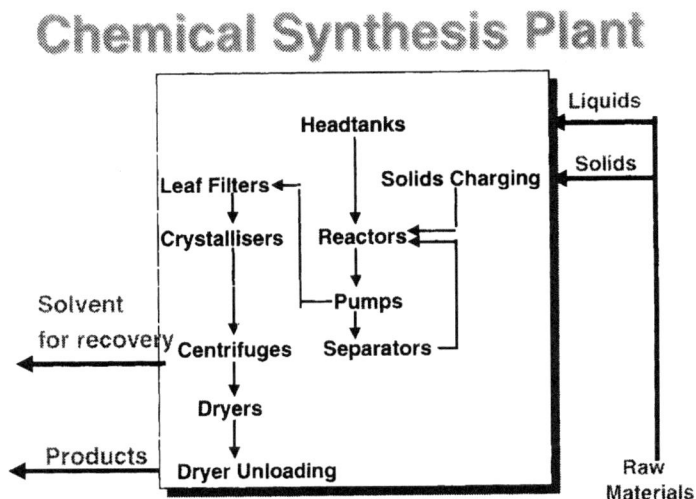

Figure 1 *Typical process flow for chemical synthesis*

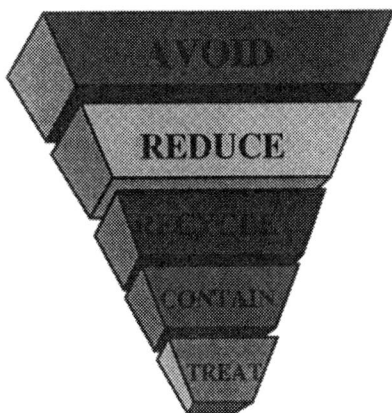

A key element in the site master plan was the integration of modern environmental management principles

Figure 2 *Hierarchy of preferred options for handling waste*

3 SUMMARY OF THE PROCESSES DEVELOPED

The recovery of the waste streams was complex, since a series of azeotropes had to be separated. Different alternatives were simulated and initial cost estimates were made by computer simulation alone. The first simulations were based only on the physical properties incorporated in the software data bank. In a second step additional physical properties {mostly liquid liquid equilibrium (LLE) data} were measured in order to increase the accuracy of the simulation of the most critical steps. First screening experiments of pervaporation to eliminate water and polar impurities such as methanol and ethanol from the tetrahydrofuran (THF) mixtures were stopped early, as it appeared that the alternatives based on counter current extraction (CCE) and rectification alone were less expensive and probably more robust. The most promising processes were piloted. The pilot experiments allowed confirmation of the results of the simulations and allowed the simulations to be updated to reflect the pilot results. A large part of the work during the pilot experiments was to verify the behaviour of further impurities contaminating the solvents, which had not been taken into account in the first screening. All impurity substances had to be purged efficiently, so that they would not accumulate after repeated recoveries of the solvents.

The sizing of the CCE-columns depended strongly on the pilot experiments. The profiles of the CCE stirrers and chamber openings had to be optimised to guarantee the desired efficiency. Finally the solvents recovered during the pilot experiments were tested in the laboratory for use in the different synthesis steps.

3.1 Process 1: Recovery of a Tetrahydrofuran/ Hexane/ isoPropylalcohol-Mixture

3.1.1 Process 1 – Inputs. The average concentration of the solvent mixture for recovery is summarized in Table 1.

3.1.2 Process 1 –Process Description. Only THF was recovered. The THF/ hexane and THF/ water azeotropes had to be separated. The rectification of THF and isopropylalcohol is difficult and requires a large number of theoretical stages. Also the concentration of THF had to be reduced to less than 8 kg/day in the wastewater stream due to its poor biodegradability. A CCE-column (111) was used to extract THF from the non polar impurities (hexane and heptane) into the aqueous phase. The LLE-data of the process are well documented[1] and the influence of isopropylalcohol could be verified by additional LLE-measurements. A column with 5 theoretical stages was used to extract THF from the hexane stream. Since the loading, interfacial tension, and the density difference between the organic and aqueous phases vary considerably as a function of the height of the CCE-column, the internals of the stirred column[2] had to be varied to ensure optimal efficiency (Figure 3).

Table 1 *Average Composition of the Solvents to be Recovered in Process 1*

	Composition w/w %
Tetrahydrofuran	67.58
Isopropylalcohol	9.81
Water	6.44
Hexane	14.74
Heptane	1.43

PROCESS 1 : Recovery of Tetrahydrofuran (THF)

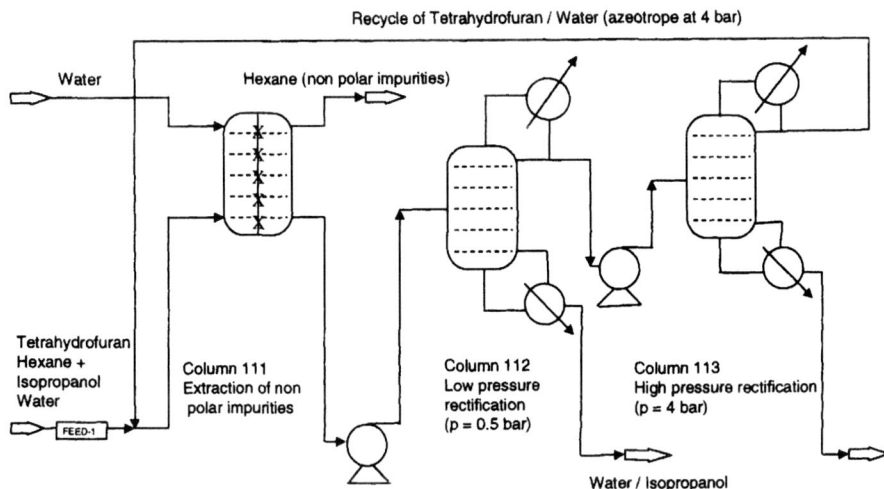

Figure 3 *Process flow diagram for Process 1*

A two pressure rectification system was used to separate water from THF. The composition of the THF/ water azeotrope varies considerably with the pressure (see Figure 4). By operating the first column (112) at 0.5 bar, the water/ THF-azeotrope with 4% (weight) water was obtained at the top of the column. The aqueous phase was THF-free and was sent to the waste water treatment plant. The distillate was sent to a high pressure column (113) operated at 4 bar, where THF with less than 0.1% weight water was the bottom product and the THF/ water azeotrope with 10% water was the distillate, which was recycled in the feed to the CCE-column 111.

Column 112 was also used to eliminate isopropylalcohol in the aqueous bottom product. Since isopropylalcohol is more polar than THF, the presence of water simplifies the separation. The relative volatility of THF/ isopropylalcohol (IPK) is increased in the presence of water (see Figure 5). An accumulation of isopropylalcohol was observed in the lowest fourth of the column. The role of the control system of the column was to keep the isopropylalcohol accumulation in the lower fourth of the column by controlling the temperature on stage 17 (see Figure 6).

3.1.3 Process 1 - Output. The recovered THF from Process 1 was greater than 99%w/w THF and contained less than 0.1%w/w of each of the main impurities. This quality was acceptable for reuse in the production of the active drug substance.

3.2 Process 2: Recovery of a THF/ Ethylacetate Mixture

3.2.1 Process 2 - Inputs. A mixture of THF/ ethylacetate contaminated by methanol, ethanol, water, and methylacetate had to be recovered. The average concentrations of the raw material are summarized in Table 2.

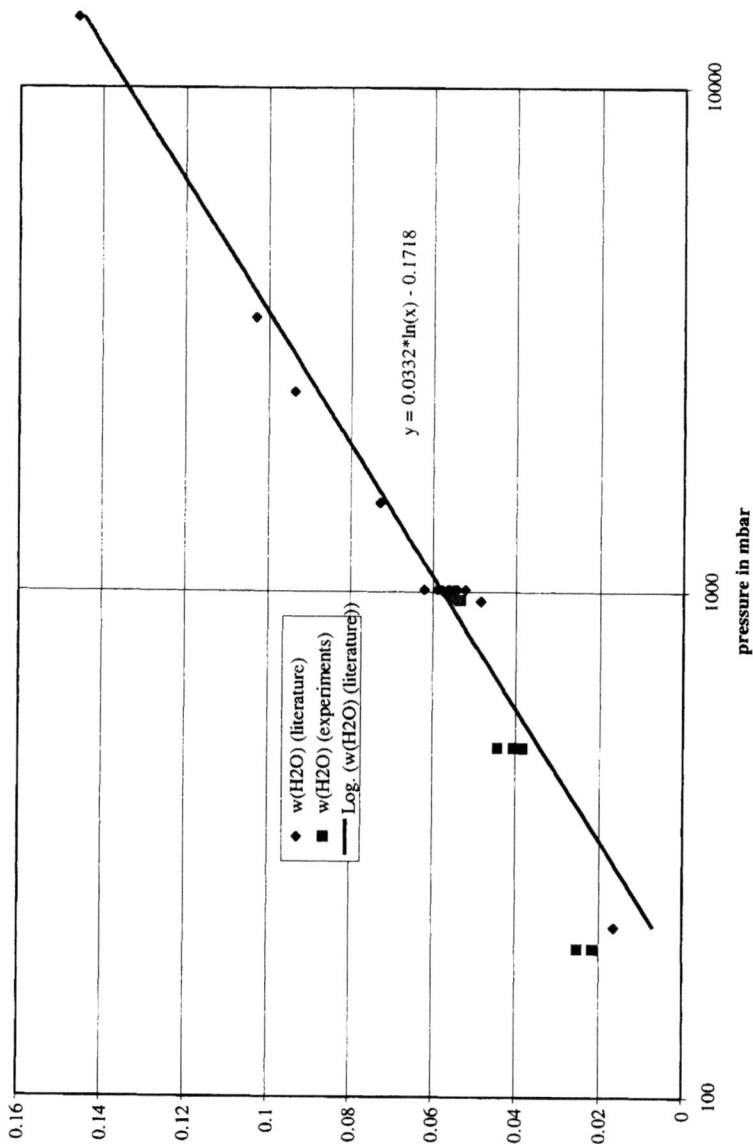

Figure 4 *Composition of the THF/water azeotrope as a function of the pressure (mass fraction water)*

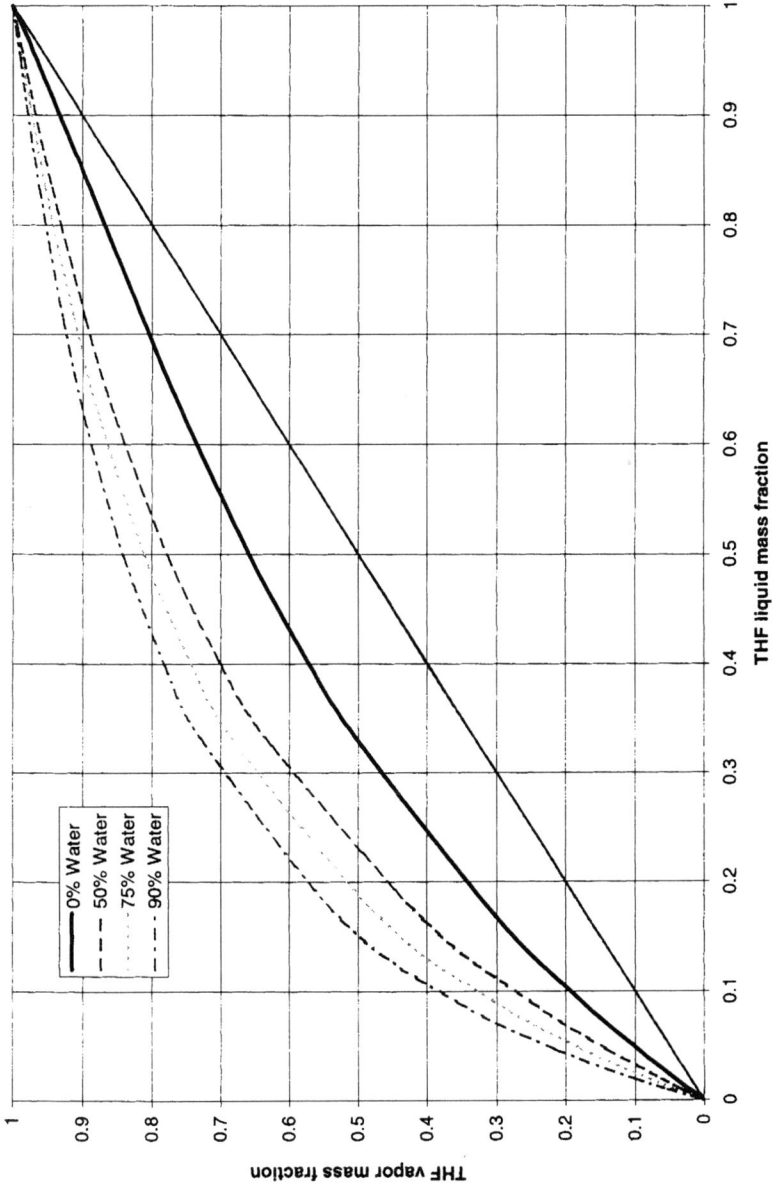

Figure 5 *xy-Diagram of THF/isopropylalcohol in the presence of 0%, 50%, 70%, and 90% water*

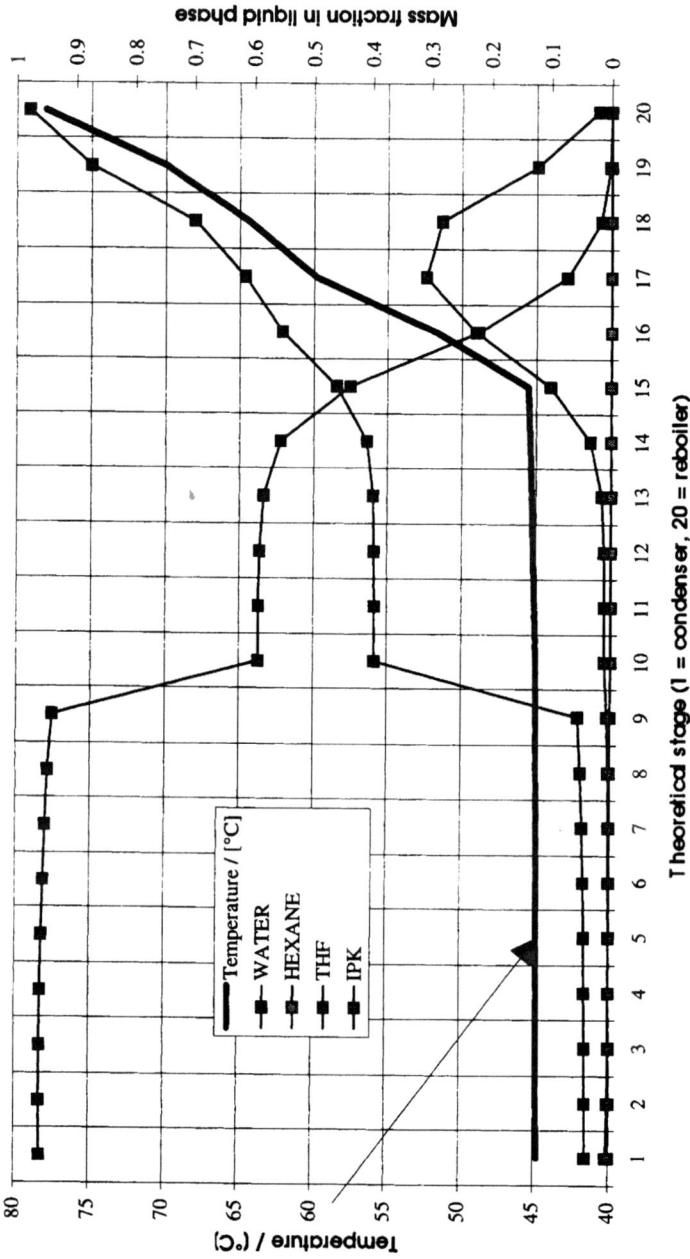

Figure 6 *Concentration profiles in the low pressure rectification column 112*

Table 2 *Composition of the THF/ Ethylacetate Mixtures*

	Composition %w/w
THF	39.26
Ethyl-acetate	39.51
Water	11.08
Methanol	8.85
Methyl-acetate	0.30
Ethanol	1
Acetic acid	trace

3.2.2 Process 2 – Process Description. The impurities in the raw material form azeotropes with tetrahydrofuran and ethylacetate. All the azeotropes had to be separated by a combination of counter current extraction and rectification. The aim was to recover ethylacetate and THF. The following major problems had to be solved by a solvent recovery unit: 1) separate the THF/ methanol and the THF/ ethanol azeotropes, 2) dewater the THF and ethylacetate (azeotropes), 3) separate THF (Atmospheric boiling point (T_B) = 65.7°C) from ethylacetate (T_B= 77°C) and methylacetate (T_B = 57.1°C).

Due to the increased difficulty of separating THF and ethylacetate in the presence's of water, it was necessary to dewater the THF/ ethylacetate-mixture before the rectification. The ethylacetate/ THF rectification column (119) was designed for 60 theoretical stages and a reflux ratio of 15. The column was operated at 0.5 bar since the VLE-data are slightly more favorable under vacuum.[3] A further reduction of the rectification pressure was not possible since water was used for condensing the distillate.

A CCE-column (116) and 2 rectification columns (117 and 118) were necessary to eliminate methanol (MY), ethanol, methylacetate, water as well as further high boiling impurities such as acetic acid before the final rectification of THF and ethylacetate in column (119) (Figure 7).

The polar components such as methanol, ethanol, and acetic acid were separated in the counter current extraction column 116. A column with 70 stirred chambers was used. Decane and water were flowing in counter current directions. Decane had to be used to ensure that phase separation occurs in the whole of the column domain, and to meet the specifications for ethanol in the organic phase and THF in the aqueous product phase. Column 116 was a fractional extraction with introduction of the feed in the middle, decane at the bottom, and water at the top. Methanol and ethanol had to be separated completely (less than 0.02% weight residual concentration in the light phase), while THF was extracted practically quantitatively in the light phase (less than 0.5% weight residual concentration in the heavy phase). The internals of the stirred columns were optimized in the pilot plant (10 theoretical stages), in order to meet simultaneously both specifications in a broad domain of water / decane ratio. Once again, large variations of the interfacial tension, density difference, and flows within the columns resulted in varying the openings between the chambers as a function of the height of the column. The use of decane simplified the dewatering of the mixture. The concentration of water in the light phase depended on the content of tetrahydrofuran, ethylacetate, and decane (see Table 3).

In order to keep the water concentration sufficiently low, the decane content in the light phase at the top of column 116 was kept between 55-60% by weight. This was done by controlling the ratio of the feed/ decane stream at the entrance of 116. The next step was a stripping column (117), which was used to dewater the organic phase. By separating 10% distillate by weight, it was possible to guarantee at the bottom product

PROCESS 2 : Recovery of Ethylacetate / Tetrahydrofuran

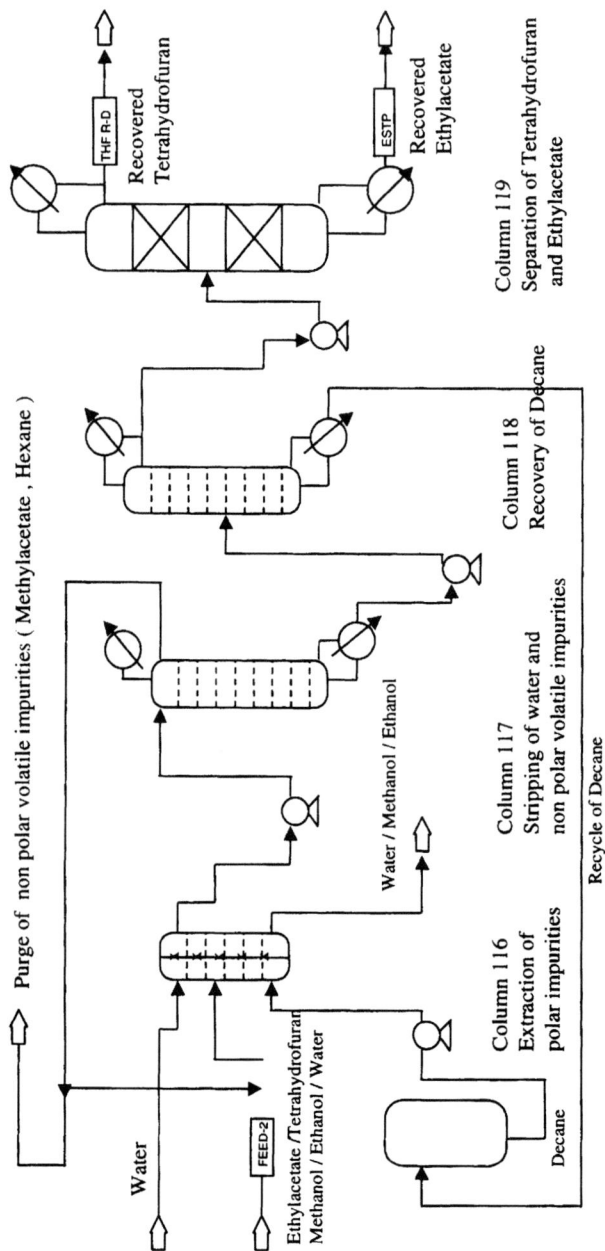

Figure 7 *Process flow diagram for Process 2*

Table 3 *Water Concentration in w/w % in the Light Product of Column 116 as a Function of the Decane/THF/Ethylacetate-Content*

Case	Methanol %	Ethylacetate %	THF %	Decane %	Water %
1	0.00	21.22	17.78	60.63	0.36
2	0.00	21.92	21.91	55.58	0.60
3	0.00	19.24	19.24	61.14	0.37
4	0.50	19.56	15.30	64.30	0.34
5	0.27	35.20	32.50	30.68	1.35
6	0.16	33.74	25.92	39.31	0.86
7	0.00	48.03	7.79	43.53	0.66

was completely water free. In order to optimize the dewatering efficiency, the column should be operated without reflux. At a higher reflux ratio, the ratio of the concentrations of THF and ethylacetate to decane increases. This reduced the dewatering efficiency. The stripping column must also concentrate methylacetate in the distillate. This was only possible at a reflux ratio of 5-10. On a continuous base, it was estimated that a 10% purge of the distillate by weight resulted in eliminating methylacetate with 2-3% overall loss by weight of THF and ethylacetate.

For a distillate flow > 26 kg/h the specifications were reached for water in the bottom product. For a distillate flow > 44 kg/h the specifications were reached for methylacetate (Figure 8).

The xy-diagram of THF/ water at 1 bar is represented for 0% decane and 50% decane in Figure 9. Decane shifts the azeotrope sufficiently to allow dewatering of the light phase from the CCE-column 116.

The third column (118) was a simple rectification column in which decane was separated from THF/ ethylacetate. Decane was recycled into the extraction column 116. Compared to different alternatives, which were simulated, this process has the following advantages. Water was eliminated from the ethylacetate/ THF-mixtures before their rectification. This approach takes advantage of the fact that the VLE-data of ethylacetate/ THF are more favorable than the ones of ethylacetate/ THF/ water. The counter current extraction with decane allows an efficient separation of the polar impurities such as methanol, ethanol, and acetic acid. Furthermore decane eliminated the water from the recovered solvent mixture (extractive rectification in column 117). Methylacetate posed a further problem and a rectification column was necessary to separate it from THF. The stripping column 117 combined the dewatering and the elimination of methylacetate.

3.2.3 Process 2 - Output. The recovered THF and ethylacetate from Process 2 were greater than 99%w/w and contained less than 0.1%w/w of each of the main impurities. This quality was acceptable for reuse in the production of the active drug substance.

3.3 Process 3: Dewatering of Ethylacetate

Column 114 was a dewatering column for a solvent, which formed a two-phase azeotrope with water. The azeotrope was obtained at the top of the column. The aqueous phase was decanted and the organic phase was refluxed at the top of the column. The optimal design of such a column implied only a stripping section since the feed was normally saturated with water (composition equal to the one of the organic phase, which is

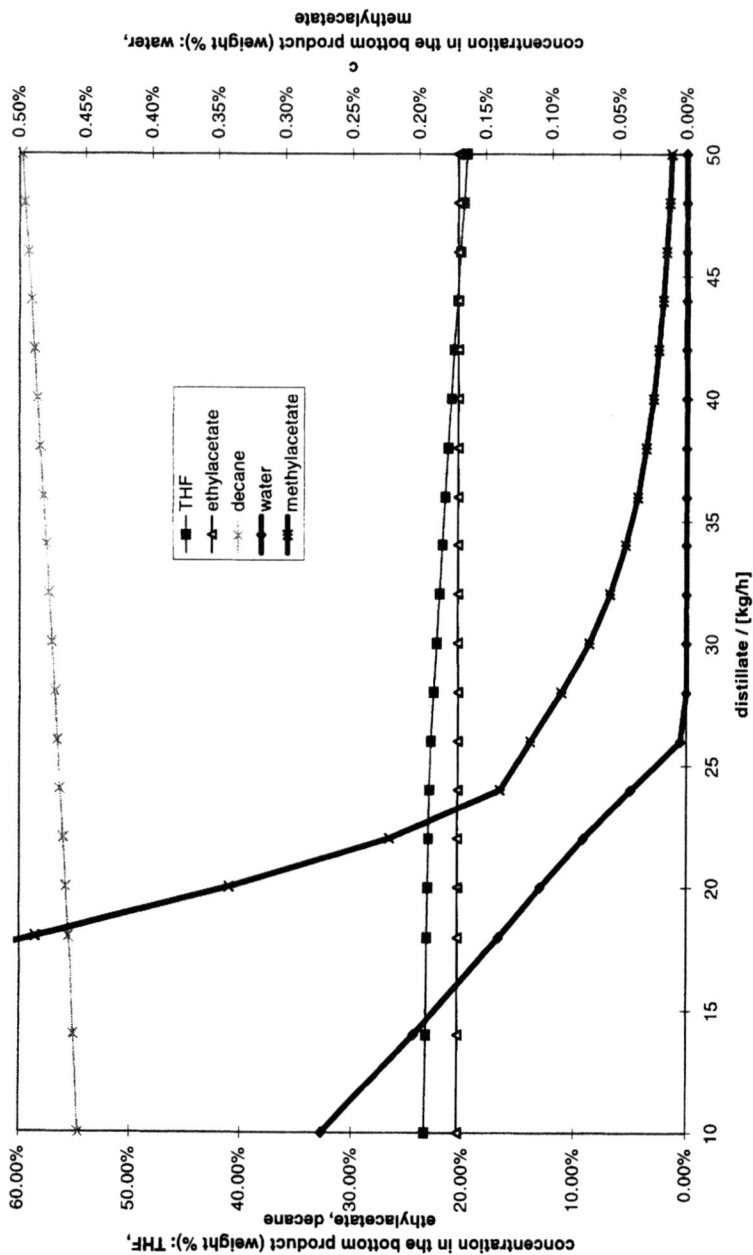

Figure 8 *Composition of the bottom product of column 117 as a function of the distillate stream (Reflux Rate = 10). The specifications are < 0.02% for water and < 0.02% for methylacetate by weight.*

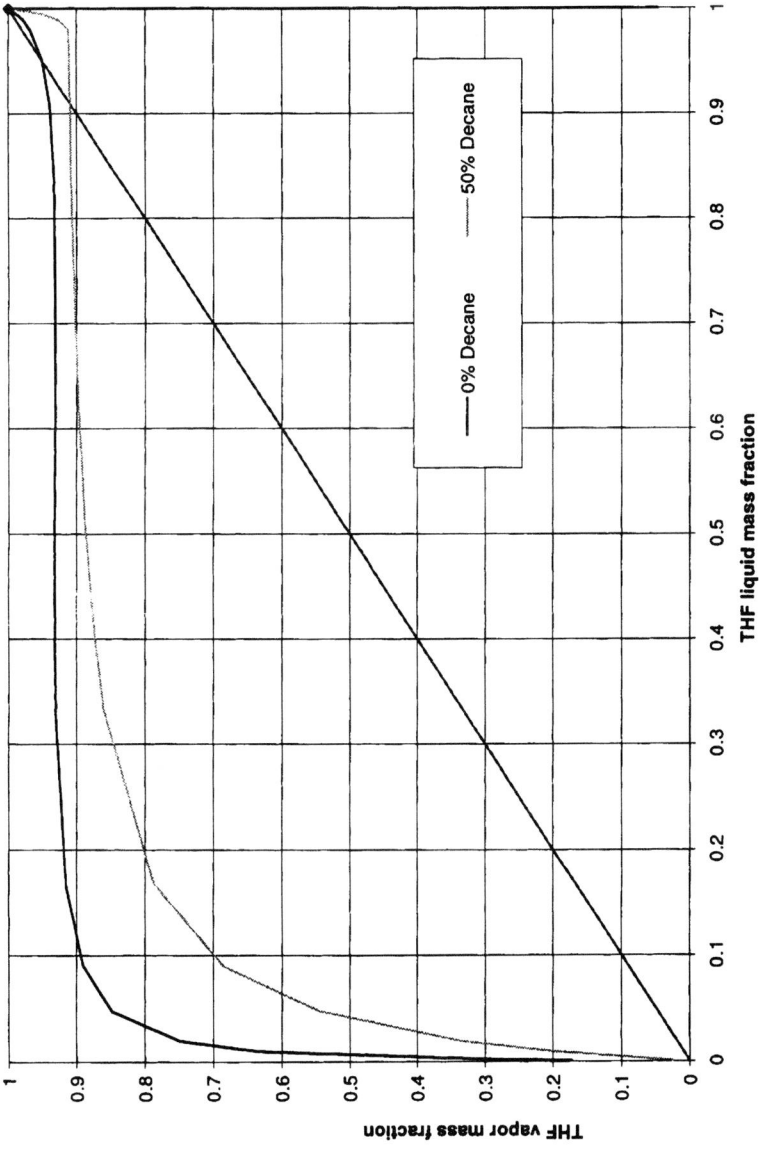

Figure 9 *Shift of the THF/water azeotrope after addition of n-decane*

PROCESS 3 : Recovery of Ethylacetate (ESTP)

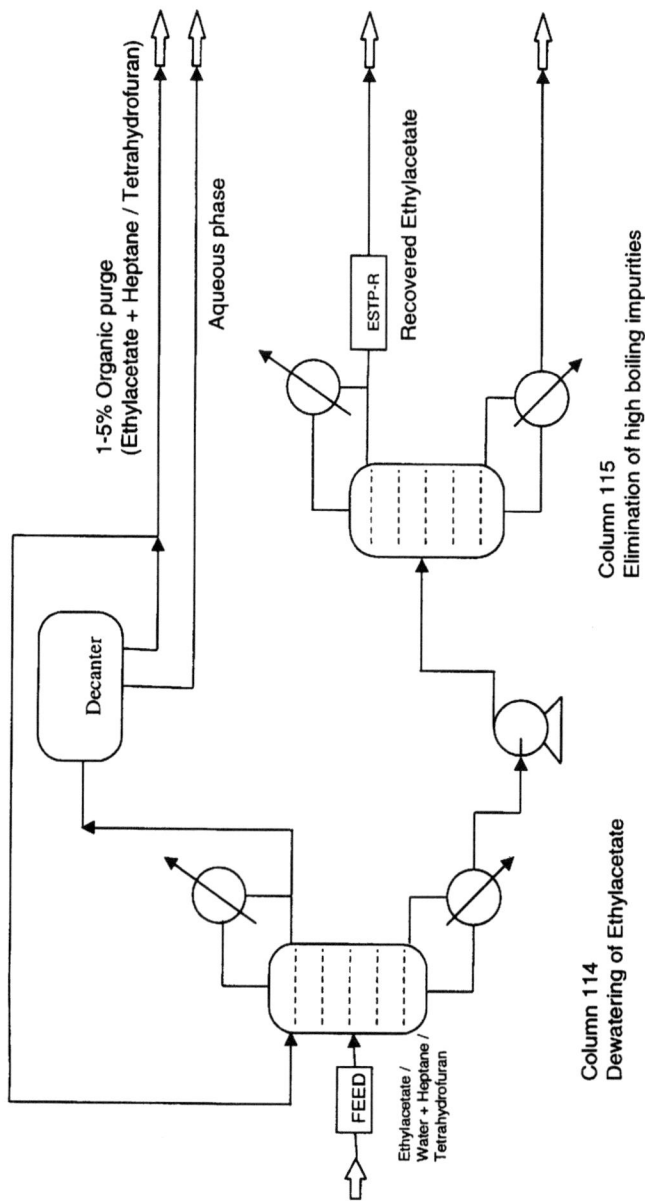

Figure 10 *Process flow diagram Process 3*

refluxed at the top of the column). The system had however to be modified to permit an efficient separation of traces of heptane and THF. A rectifying section with approximately 10 theoretical stages was added in order to concentrate heptane (forming an azeotrope with ethylacetate) and THF at the top of the column. An organic purge eliminated more than 90% of these impurities by weight. An additional rectification column (115) was used to eliminate high boilers. The key component taken to design the column was acetic acid, resulting in a column with 15 theoretical plates operated at a reflux ratio of 1.5 (Figure 10).

4 GENERAL APPLICABILITY

Complex multiphase process waste streams are generated in the production of active substances in the chemical industry. A number of standard types of chemistry reactions currently carried out in the pharmaceutical industry would generate complex waste solvent mixtures containing tetrahydrofuran, hexane, alcohols and water.

Examples of such reactions are as follows:

- a) Grignard reactions.
- b) Reductions with metal hydrides such as sodium borhydride and lithium aluminium hydride.
- c) Hydroborations.
- d) Simmons - Smiths reactions.
- e) A significant part of the aldol or Claisen condensations.
- f) Low temperature reactions involving strong bases like sodium hydride, butyllithium and hexyllithium.

Other industries carrying out organic synthesis such as the agrochemical industry or a part of the fine chemical industry use similar reactions.

Common to most of these reactions is that they are generally done in the absence of water and that the product mixtures are quenched in water or alcohols. The solvent mixtures are further contaminated with by-products of the reaction as well as solvents which are either used for dissolving the reagents or later during the work-up of the reaction mixture. In many cases a recovery unit similar to the one outlined in this paper would permit to recover and reuse the solvents.

The commercial viability needs to be considered on a case by case basis. The commercial viability will depend on factors such as the volume of spent solvent and the potential value of the solvent recoverable from the waste stream.

References

1. J. Lampa, J. Matous, J. Novak and J. Pick, *Collect. Czech. Chem. Commun.*, 1980, **45**, 1159.
2. U. Bühlmann and A. Mögli, 'The Kühni Extraction Column', in 'Handbook of Solvent Extraction', (Edited by T. C. Lo, M. Baird and C. Hanson), Wiley, New York, 1983, 441-447.
3. H. S. Wu and S. I. Sandler, 'Vapor-Liquid Equilibria of Tetrahydrofuran Systems', *J. Chem. Eng. Data*, 1988, **33**, 157.

Subject Index

CPSIA information can be obtained at www.ICGtesting.com
Printed in the USA
LVOW111521010712

288401LV00007B/62/A

9 780854 048854